INTEGRATING VARIABLE RENEWABLE ENERGY IN ELECTRICITY MARKETS

INTERNATIONAL EXPERIENCES

RENEWABLE ENERGY: RESEARCH, DEVELOPMENT AND POLICIES

Additional books in this series can be found on Nova's website
under the Series tab.

Additional E-books in this series can be found on Nova's website
under the E-book tab.

INTEGRATING VARIABLE RENEWABLE ENERGY IN ELECTRICITY MARKETS

INTERNATIONAL EXPERIENCES

SOPHIA B. TAYLOR
AND
PATRICK R. YOUNG
EDITORS

New York

Copyright © 2013 by Nova Science Publishers, Inc.

For permission to use material from this book please contact us:
Telephone 631-231-7269; Fax 631-231-8175
Web Site: http://www.novapublishers.com

NOTICE TO THE READER

The Publisher has taken reasonable care in the preparation of this book, but makes no expressed or implied warranty of any kind and assumes no responsibility for any errors or omissions. No liability is assumed for incidental or consequential damages in connection with or arising out of information contained in this book. The Publisher shall not be liable for any special, consequential, or exemplary damages resulting, in whole or in part, from the readers' use of, or reliance upon, this material. Any parts of this book based on government reports are so indicated and copyright is claimed for those parts to the extent applicable to compilations of such works.

Independent verification should be sought for any data, advice or recommendations contained in this book. In addition, no responsibility is assumed by the publisher for any injury and/or damage to persons or property arising from any methods, products, instructions, ideas or otherwise contained in this publication.

This publication is designed to provide accurate and authoritative information with regard to the subject matter covered herein. It is sold with the clear understanding that the Publisher is not engaged in rendering legal or any other professional services. If legal or any other expert assistance is required, the services of a competent person should be sought. FROM A DECLARATION OF PARTICIPANTS JOINTLY ADOPTED BY A COMMITTEE OF THE AMERICAN BAR ASSOCIATION AND A COMMITTEE OF PUBLISHERS.

Additional color graphics may be available in the e-book version of this book.

Library of Congress Cataloging-in-Publication Data

ISBN: 978-1-62808-572-3

Published by Nova Science Publishers, Inc. † New York

CONTENTS

PREFACE

Chapter 1 – Many countries—reflecting very different geographies, markets, and power systems—are successfully managing high levels of variable renewable energy on the electric grid, including that from wind and solar energy. This study documents the diverse approaches to effective integration of variable renewable energy among six countries—Australia (South Australia), Denmark, Germany, Ireland, Spain, and the United States (Colorado and Texas)—and summarizes policy best practices that energy ministers and other stakeholders can pursue to ensure that electricity markets and power systems can effectively coevolve with increasing penetrations of variable renewable energy. There is no one-size-fits-all approach; each country has crafted its own combination of policies, market designs, and system operations to achieve the system reliability and flexibility needed to successfully integrate renewables. Notwithstanding this diversity, the approaches all coalesce around five strategic areas: lead public engagement, particularly for new transmission; coordinate and integrate planning; develop rules for market evolution that enable system flexibility; expand access to diverse resources and geographic footprint of operations; and improve system operations. The study also emphatically underscores the value of countries sharing their experiences. The more diverse and robust the experience base from which a country can draw, the more likely that it will be able to implement an appropriate, optimized, and system-wide approach.

Chapter 2 – This project examines renewable energy deployment in the United States using a version of the Global Change Assessment Model (GCAM) with a detailed representation of renewables, the GCAM-RE. Electricity generation was modeled in four generation segments and 12-subregions. This level of regional and sector detail allows a more explicit representation of renewable energy generation. Wind, solar thermal power, and central solar PV plants are implemented in explicit resource classes with new intermittency parameterizations appropriate for each technology. A scenario analysis examines a range of assumptions for technology characteristics, climate policy, and long distance transmission. We find that renewable generation levels grow over the century in all scenarios. As expected, renewable generation increases with lower renewable technology costs, more stringent climate policy, and if alternative low-carbon technologies are not available. The availability of long distance transmission lowers policy costs and changes the renewable generation mix.

In: Integrating Variable Renewable Energy ...   ISBN: 978-1-62808-572-3
Editors: S.B. Taylor and P.R. Young   © 2013 Nova Science Publishers, Inc.

Chapter 1

INTEGRATING VARIABLE RENEWABLE ENERGY IN ELECTRIC POWER MARKETS: BEST PRACTICES FROM INTERNATIONAL EXPERIENCE[*]

*Jaquelin Cochran, Lori Bird, Jenny Heeter
and Douglas J. Arent*

ABSTRACT

Many countries—reflecting very different geographies, markets, and
power systems—are successfully managing high levels of variable
renewable energy on the electric grid, including that from wind and solar
energy. This study documents the diverse approaches to effective
integration of variable renewable energy among six countries—Australia
(South Australia), Denmark, Germany, Ireland, Spain, and the United
States (Colorado and Texas)—and summarizes policy best practices that
energy ministers and other stakeholders can pursue to ensure that
electricity markets and power systems can effectively coevolve with
increasing penetrations of variable renewable energy. There is no one-
size-fits-all approach; each country has crafted its own combination of
policies, market designs, and system operations to achieve the system
reliability and flexibility needed to successfully integrate renewables.

---

[*] This report, NREL/TP-6A00-53732, was released by the National Renewable Energy Laboratory, April 2012.

Notwithstanding this diversity, the approaches all coalesce around five strategic areas: lead public engagement, particularly for new transmission; coordinate and integrate planning; develop rules for market evolution that enable system flexibility; expand access to diverse resources and geographic footprint of operations; and improve system operations. The study also emphatically underscores the value of countries sharing their experiences. The more diverse and robust the experience base from which a country can draw, the more likely that it will be able to implement an appropriate, optimized, and system-wide approach.

LIST OF ACRONYMS AND ABBREVIATIONS

AC	alternating current
AEMO	Australian Energy Market Operator
AWEA	American Wind Energy Association
BNetzA	*Bundesnetzagentur Startseite* (Federal Network Agency for Electricity, Gas, Telecommunications, Post and Railways, Germany)
CAISO	California ISO (independent system operator)
CEM	Clean Energy Ministerial
CHP	combined heat and power
CREZ	competitive renewable energy zone
DCENR	Department of Communications, Energy and Natural Resources (Ireland)
DLR	*Deutsches Zentrum für Luft- und Raumfahrt* (Germany)
DNSP	Distribution Network Service Providers
DOE	U.S. Department of Energy
DPS	Drift *Planlaegnings* (Operational Planning) System (Germany)
DS3	Delivering a Secure Sustainable Electricity System
DWEA	Danish Wind Energy Association
EIM	energy imbalance market
ENTSO-E	European Network of Transmission System Operators for Electricity
ERCOT	Electric Reliability Council of Texas
ESIPC	Electricity Supply Industry Planning Council (South Australia)
EUR	euro

EWIS	European Wind Integration Study
EWITS	Eastern Wind Integration and Transmission Study
FERC	Federal Energy Regulatory Commission
FIT	feed-in tariff
GEMAS	maximum admissible wind generation (Spain)
GIVAR	Grid Integration of Variable Resources (IEA)
GW	gigawatts
GWC	Global Weather Corporation
GWEC	Global Wind Energy Council
GWh	gigawatt-hour
HVDC	high voltage, direct current
Hz	hertz
IDAE	Institute for Energy Diversification and Saving
IEA	International Energy Agency
IEEE	Institute of Electrical and Electronics Engineers
IPCC	Intergovernmental Panel on Climate Change
ISO	independent system operator
ITVC	Interim Tight Volume Market Coupling
IWEA	Ireland Wind Energy Association
KfW	*Kreditanstalt fuer Wiederaufbau* (Germany)
LaaR	Load Acting as a Resource
LMP	locational marginal pricing
MWh	megawatt-hour
n.d.	not dated
NCAR	National Center for Atmospheric Research
NDP	National Development Plan (Australia)
NEL	National Electricity Law (Australia)
NEM	National Electricity Market (Australia)
NERC	North American Electric Reliability Corporation
NOIS	Nordic Operational Information System
NREL	National Renewable Energy Laboratory
NTNDP	National Transmission Network Development Plan (Australia)
OCGT	open-cycle gas-turbine
PUCT	Public Utilities Commission of Texas
R&D&I	research and development and innovation
REE	*Red Eléctrica de España* (Spain)
REFIT	Renewable Energy Feed-in Tariff
REN21	Renewable Energy Policy Network for the 21st Century

RETD	Renewable Energy Technology Deployment
RETI	Renewable Energy Transmission Initiative (California)
RIT-t	Regulatory Investment Test – transmission
RTO	regional transmission organization
SEM	Single Electricity Market
SEMO	Single Electricity Market Operator
SKM	Sinclair Knight Merz
SONI	System Operators Northern Ireland
TNSP	Transmission Network Service Providers
TWh	terawatt-hour
TSO	transmission system operator
UNIDO	United Nations Industrial Development Organization
WSAT	Wind Security Assessment Tool
WECC	Western Electricity Coordinating Council
WFES	World Future Energy Summit

EXECUTIVE SUMMARY

Many countries—reflecting very different geographies, markets, and power systems—are successfully managing high levels of variable renewable energy (RE) on the grid, such as from wind and solar energy. This study documents the diverse approaches to effective integration of variable RE among six countries—Australia (South Australia), Denmark, Germany, Ireland, Spain, and the United States (Colorado and Texas)—and summarizes policy best practices that energy ministers and other stakeholders can pursue to ensure that electricity markets and power systems can effectively coevolve with increasing penetrations of variable RE.

The cases studied reveal there is no one-size-fits-all approach; each country has crafted its own combination of policies, market designs, and system operations to achieve the system reliability and flexibility needed to successfully integrate variable RE. Notwithstanding this diversity, the approaches coalesce around five strategic areas of intervention.

A. Lead Public Engagement, Particularly for New Transmission

Installing transmission that may be required to accommodate new RE can be challenging; stakeholders may express concerns over land use changes, environmental damage, decreased property values, or health concerns. A foundational component of siting new transmission is public engagement—a two-way exchange of information.

> Action to Improve Public Support: lead public engagement to communicate to the public why new transmission is essential
> Examples: public stakeholder exchange in Texas and Germany; burying of high-voltage grid in Denmark

B. Coordinate and Integrate Planning

Planning comprises an inherently complex set of activities that are undertaken by multiple groups and jurisdictions for a given power system. Variable RE can be accommodated by integrating the planning of generation, transmission, and system performance; ensuring institutions and markets are designed to enable access to physical capacity; and building from local and regional planning to better integrate and coordinate information across jurisdictions.

> Actions to Improve Planning: share best practices and guidelines for adapting advanced planning capabilities; support capacity of institutions to increase integration, complexity, and coordination of—and stakeholder participation in—planning; and provide vision for how to move from analyses and recommendations to actions
> Examples: market-based guided development in Australia; centralized planning in Texas

C. Develop Rules for Market Evolution that Enable System Flexibility

Higher penetrations of variable RE require increased flexibility from the power system. Flexible generation can be encouraged by implementing sub-hourly scheduling and dispatch intervals (5- or 15-minute) and shorter gate closure periods; establishing capacity and other ancillary services markets; and developing zonal or nodal pricing.

Markets that encourage flexible storage and demand response can support the integration of variable RE through load shifting, balancing, and frequency and regulation.

Actions to Support Flexibility: lead the development and innovation of market designs; encourage market operators to adopt rules to improve system efficiency; and play a leading role in negotiating a framework for integration that optimizes flexibility across regions

Examples: market design that encourages system flexibility through expanded power markets, fast market design, combined heat and power (CHP), and negative prices in Denmark; National Electricity Market operation in Australia; encouragement of storage in Germany; demand response participation in Texas

D. Expand Access to Diverse Resources and Geographic Footprint of Operations

Integration studies have consistently found that expanding access to diverse resources aids the integration of high penetrations of variable RE. This can be achieved in two ways: enlarging effective balancing areas, and diversifying the location and types of RE generation. By enlarging balancing areas, the relative variability and uncertainty in both the load and RE generation will be lowered. Greater geographic distribution of renewable resources reduces the variability of RE because weather patterns are less correlated, reducing the magnitude of output changes. Regional market pricing, RE zones, and hybrid market solutions that allow reserve sharing are some of the approaches to encourage diversity.

Actions to Expand Diversity: support the study and evaluation of options to diversify resources and enlarge balancing areas, and convene stakeholder discussions to overcome institutional challenges in merging or increasing cooperation among balancing areas

Examples: expanding regional integration in Ireland; proposed Energy Imbalance Market in the Western U.S.

E. Improve System Operations

Integrating advanced forecasting techniques into fast market operations, the control room, and other standard operating practices can help predict the amount of RE available to the system. Grid codes—rules that govern how power plants connect to and support the grid—help ensure that variable RE is compatible with, and can help contribute to, the stability of, the power grid. A necessary first step is to evaluate existing rules to determine whether new approaches to planning, design, and operation are needed for high penetrations of variable RE.

Actions to Improve System Operations: support development of national or regional forecasting systems and work with regulatory commissions to evaluate model grid codes, recommend changes, and implement recommendations

Examples: market use of advanced forecasting in Australia; system operator use of multiple and advanced forecasting in Denmark; creation of the innovative Control Centre for Renewable Energies in Spain

This study emphatically underscores the value of countries sharing their experiences. Any country's ability to successfully integrate variable RE depends on a wide array of factors: technical requirements, resource options, planning processes, market rules, policies and regulations, and institutional and human capacity. The more diverse and robust the experience base from which a country can draw, the more likely that it will be able to implement an appropriate, optimized, and system-wide approach. This is as true for countries in the early stages of RE integration as it is for countries that have already had significant success. Going forward, successful RE integration will thus depend upon the ability to maintain a broad ecosystem perspective, to organize and make available the wealth of experiences, and to ensure that there is always a clear path from analysis to enactment.

INTRODUCTION

Economic, environmental, and security concerns associated with conventional fuel supplies have strengthened support for clean energy technologies among governments and the private sector on a global scale; yet, questions about how to effectively integrate large amounts of variable renewable energy (RE)[1] generation persist. RE accounted for nearly half the estimated 194 gigawatts of new global capacity in 2010—an investment equal to $211 billion (REN21 2011). Variable renewables, in particular, have achieved significant penetration in many countries, and issues associated with grid integration are increasingly gaining attention among a broad range of stakeholders.

The depth of experience in various countries—situated in diverse geographical and market contexts—provides insights for decision makers interested in increasing the penetration of variable RE into the power sector.

This study documents the diverse approaches to effective integration among six countries, and it summarizes policy best practices that energy ministers and other stakeholders can pursue to ensure that electricity markets and power systems can effectively coevolve with increasing penetrations of variable RE.

APPROACH

Many countries—reflecting very different geographies, markets, and power systems—are demonstrating success in managing high levels of variable RE on the grid, such as from wind and solar. The Clean Energy Ministerial,[2] which seeks to advance clean energy globally, identified the value in learning from this diverse set of experiences, so that these lessons can be applied elsewhere. The energy ministers participating in the Clean Energy Ministerial requested a review of the approaches taken by these countries, lessons that can be learned, and actions that energy ministers and other stakeholders can take to create supportive markets, institutions, and power systems.

The cases selected for this study—Australia (South Australia), Denmark, Germany, Ireland, Spain, and the United States (Western Region: Colorado and Texas)—all have relatively high penetrations of RE but reflect different power system and market characteristics.

BACKGROUND

Variable RE presents an opportunity to exploit local renewable resources and reduce emissions. However, variable RE is often perceived as incompatible with base load needs and a secure electricity grid because of its inherent uncertainty in availability. Commonly cited concerns include technical, financial, geographical, and institutional concerns.

Technical Concerns

Higher penetrations of variable RE require increased flexibility from the power system to manage the variability and uncertainty of the generation.

Also, although many RE generators now have technologies that support grid reliability,[3] a misperception persists that they cannot do so.

Financial Concerns

Variable RE, with low to zero marginal costs, can impact the wholesale price of power and in turn the revenues and returns on investments for other generators. Also, timescale misalignment between transmission and generation planning and construction necessitates financial risk (e.g., RE investors may build generation in advance of transmission, and thus face revenue risk, or alternatively generators may have to pay for transmission in advance of generation coming on line). Finally, RE curtailments, which might increase in frequency at higher penetrations of RE, could cause unexpected revenue shortfalls.

Geographical Concerns

Variable RE is not necessarily generated where load is. Adding new variable RE may increase both transmission bottlenecks and the need for new transmission lines to remote areas.

Institutional Concerns

Regulatory and legislative barriers reduce access to optimal solutions. Compounding such difficulties is the need to coordinate across many jurisdictions and agencies (e.g., markets, balancing areas, planning entities, forecasters) to comprehensively address RE integration. Finally, integrating RE may require an educational or cultural shift among stakeholders to address the need for operational changes and increased flexibility needed to manage high penetrations of renewables on the grid.

Existing studies and tasks address many of these concerns:

- The IEA Grid Integration of Variable Resources (GIVAR) project seeks to better understand the technical and market characteristics of a power system that facilitates integration of variable RE. For example, the Phase 2 report, *Harnessing Variable Renewables* (IEA 2011), proposes a tool to assess how much renewable energy can be added to existing systems.

- Task 25 of the IEA Wind Implementing Agreement establishes and shares best practices for increasing wind energy penetration.
- The Intergovernmental Panel on Climate Change (IPCC) Special Report on Renewable Energy (IPCC 2011) includes an assessment on integrating large amounts of renewable energy and recommended policy best practices.
- The North American Electric Reliability Corporation (NERC)'s *Accommodating High Levels of Variable Generation* (NERC 2009) outlines the state of knowledge and further research needed to integrate variable RE.
- Various Institute of Electrical and Electronics Engineers (IEEE) power and energy task forces investigate technical concerns about integration (ongoing).
- Integration studies (see Text Box 7) evaluate opportunities, actions needed, and associated costs to integrate wind and solar energy to the grid.
- Alstom Grid Inc.'s Strategies and Decision Support Systems for Integrating Variable Energy Resources in Control Centers for Reliable Grid Operations (Jones 2011) identifies global trends in how system operations are responding to increased wind penetration, and their perspectives on effective approaches and future needs.

This report draws from both these studies and the compendium of case studies selected for this study (Appendices A-G) to synthesize lessons on effective policies, regulations, planning, and practices to encourage integration of significant RE penetration into electricity systems. Text Box 1 explains terms related to the integration of variable RE used in this study.

Text Box 1. Terms Used in this Study

- **Balancing area** includes all generation, transmission, and load within a metered area. The balancing authority maintains the balance (a period spanning minutes to days) of demand, generation, and net flows of power to adjacent balancing areas.
- **Curtailment,** or restriction on output of variable RE generation, can result in lost revenue for generators, but is sometimes used to support system flexibility.

Curtailment can provide more time for other plants on the system to ramp down if there is a sudden imbalance of high supply and low demand. Curtailment can also be used to require an RE generator to generate at reduced levels so that it can ramp up quickly to balance a system.

- **Fault ride-through capability** refers to the ability of generators to operate through low and high voltages or low and high frequency events. Historically variable RE generators, such as wind generators, were required to disconnect when, for example, the voltage dropped at the wind turbine due to a fault in the grid. With increasing penetrations of variable RE, the power system is more reliable as generators can continue to operate and support the power system in these circumstances. Thus, variable RE generators are increasingly required to have fault ride-through capabilities.
- **Flexibility** refers to how quickly a resource can respond to expected or unexpected changes in demand or supply. Historically, flexible resources have been pumped storage, hydropower, and gas peaking plants. The amount of system flexibility depends not only on technical aspects (e.g., extent of interconnection, storage, forecast errors, and speed of response of resources) but also on institutional and market aspects (e.g., market rules that allow for short gate closure times, demand response programs). Net flexibility refers to system flexibility minus flexibility requirements necessary for grid operation.
- **Gate closure** refers to the future time at which the market commits to deliver electricity. Gate closures that occur close to the actual delivery time (e.g., 5 or 15 minutes in advance) can help minimize the magnitude of forecast errors and associated reserves and allow for trading at potentially lower costs than power that would otherwise be required to balance day-ahead schedules.
- **Net load** refers to electricity demand minus electricity supplied by variable RE and hence the electricity that must be supplied by other resources. The concept of net load helps grid operators assess the required response from conventional energy supplies to balance the system; ramping might be smoothed when RE resources such as solar complement demand, or more likely, they might be made more severe when RE supply fluctuates independent of demand.

- **Reactive power** (also referred to its unit of measurement, "vars") helps maintain voltage in the transmission system needed to circulate "active" power. Modern wind turbines can provide this voltage control.
- **System frequency** reflects the balance between demand and supply; a decreasing frequency indicates demand has increased above supply. A stable power system does not allow frequency to deviate more than a small fraction (~0.03 Hz) from the 50 or 60 Hz target. As with fault ride-through capabilities, the power system is more reliable when variable RE generators can provide frequency response.
- **System inertia**—Inertia, the rotating energy stored in the system, helps maintain system stability against sudden frequency changes (e.g., changes that occur due to imbalances between system load and generation). Synchronous generators help maintain system inertia and therefore system frequency. Higher penetrations of RE increase the likelihood of sudden changes and could be of particular concern if new generating technologies cannot contribute to system inertia.

FIVE AREAS OF INTERVENTION TO ACCOMMODATE HIGH RE PENETRATION

Analysis of the results from the case studies conducted for this study reveals a wide range of mechanisms that can be used to accommodate high penetrations of variable RE (e.g., from new market designs to centralized planning). Nevertheless, the myriad approaches collectively suggest that governments can best enable variable RE grid integration by implementing best practices in five areas of intervention:

A. Lead public engagement, particularly for new transmission
B. Coordinate and integrate planning
C. Develop rules for market evolution that enable system flexibility
D. Expand access to diverse resources and geographic footprint of operations
E. Improve system operations.

Figure 1 illustrates, within each of these areas, when actions typically need to be implemented as a country transitions from low to high RE penetration. For each of the five areas of intervention, the following sections describe in detail the rationale, best practices, diversity of approaches as revealed through the case studies, and challenges and actions to implementing best practices. Additional details on the case studies can be found in Appendices A-G. Cochran et al. (2012) summarize this study for policymakers.

	Public Outreach	Planning	Market Rules	Expanded Access	System Operations
At LOW RE Penetrations	Involve public stakeholders in planning	Evaluate system flexibility, penetration scenarios, transmission needs, and future flexibility needs	Evaluate market design and implications for higher penetrations of RE	Assess renewable energy resources and options for encouraging geographic diversity	Build capacity of grid operator staff; review regulatory changes needed to require advanced forecasting
At MEDIUM RE Penetrations	Communicate to public why new transmission is essential	Regulatory and legislative changes needed to accommodate revised scenario planning, such as laws to support renewable energy zones (REZs)	Ensure that market design and pricing environment aligns with technical needs, such as accessing flexibility, minimizing uncertainty, and managing risk	Make necessary regulatory, market, or institutional changes	Implement grid codes to accommodate high penetrations of variable RE
At HIGH RE Penetrations		Monitor and review effectiveness of actions; revise	Ensure broad systems solutions are sought, including smart grid/demand response, storage, and complementary flexible generators		

Figure 1. Key activities in transitioning from low to high RE penetration.

A. Lead Public Engagement including Facilitating New Transmission

Rationale

High penetrations of variable RE may require expanded transmission capacity—to accommodate diverse RE locations and locations far from load, to enlarge balancing areas, to reduce nodes of transmission congestion, and to fully access flexible resources (generation, storage, and demand response).

Installing this transmission, however, is a challenge; stakeholders may express concerns over land use changes, environmental damage, decreased property values, or health concerns. Further, who should pay for transmission investments—and how these costs are allocated among ratepayers, stockholders, and others—can be an issue in some cases. Negotiating the balance between new transmission projects and the potential conflicts that arise often requires political leadership.

Text Box 2. Example of Approach to Public Engagement—Texas

Competitive renewable energy zones (CREZ) exemplify a quick process for gaining public support for new transmission. In implementing CREZ, the Public Utilities Commission of Texas (PUCT) authorized transmission improvements to serve 18.5 GW in new generation capacity—an equivalent to nearly 3,000 miles of high-voltage transmission lines. The PUCT identified transmission providers, who in turn designed each route—a 1-1.5 year approval process; some contested cases involved more than 1,000 landowners (Lasher 2011). Mandatory open houses for the public allowed developers to explain, for example, the purpose and need for the project and environmental and engineering constraints. Landowners were able to inform developers of siting considerations not available in the public record, such as locations of public landing strips and cemeteries, and to provide suggestions on unobjectionable routing locations (Reid 2012). In addition to the open houses, a multi-page questionnaire sought landowners' feedback on, for example, their understanding of the need for the project and relative importance of various factors in siting considerations. Letters sent to county commissioners, judges, farm bureaus, federal agencies, and others allowed for additional feedback, such as locations of planned schools. All stakeholders could also offer testimony in hearings, which formed the basis of the final siting decision by the PUCT.

As an example outcome of this process, the PUCT required that Oncor, which is building more than 1,000 miles of transmission, use monopole structures in places to reduce land impacts, and that 44% of routing be parallel to existing lines (Doan 2010). According to Oncor, "The grassroots attendees who started as protesters applauded when the routes were selected, which is proof that the process works" (Doan 2010).

According to the latest schedule for CREZ, all transmission is to be energized by the end of 2013, eight years after the state legislature initiated the CREZ in 2005. A few additional factors helped expedite the process:

- Most of the routing was over private lands; other locations in the United States often involve federal land, a greater range of stakeholders, and compliance with the National Environmental Policy Act.
- Initial selection of CREZ locations was based on confidential information provided by wind energy companies. ERCOT chose locations based on significant interest and did not allow citizen input into this process.
- The purpose of the CREZ-designated lines was not allowed as an issue of intervention, reducing public input to decisions on location, not need.

Best Practices

Public engagement and a two-way exchange of information are foundational components of siting new transmission; transmission developers can explain why new transmission lines are needed and the public can offer feedback on siting locations (e.g., planned locations for new schools) and other information not available in public records.

Text Box 3. Example of Approach to Facilitating New Transmission—Germany

Germany enacted two new laws to facilitate new transmission. The 2009 Power Grid Expansion Act (EnLAG) gives priority to extra-high voltage transmission projects that reduce north/south congestion. The 2011 Grid Expansion Acceleration Law (NABEG) intends to shorten planning and permission processes from the current ten years to four years by consolidating responsibilities at the federal government level and allowing for early public participation (BMU 2012).

To further facilitate transmission development, TenneT, one of four transmission system operators (TSOs) in Germany, has requested higher allowable rates of return from BnetzA (the federal network regulator) in order to attract investors.

TenneT has also proposed a joint direct current network operator to build the dc line for offshore connections (TenneT 2012). BnetzA is considering a host of reforms, including whether to remove the current two-year waiting period that grid companies face before they are able to recover costs.

> To facilitate public engagement, the TSOs in Germany have developed a website to educate the public on the decision-making process for new transmission; in-depth public consultation will be held and municipalities may be compensated for hosting new transmission corridors.

California's Renewable Energy Transmission Initiative (RETI)[4] is often cited as a model for building public support for new transmission. The objective of RETI is to facilitate siting and permitting for renewable energy generation and transmission projects. A 30-person Stakeholder Steering Committee for RETI represents diverse constituencies, including utilities, generators, state and federal officials, environmental groups, consumers, military, and American Indian Tribes. Committee members commit to engage and incorporate feedback from diverse stakeholders, work in good faith to achieve consensus on key issues, and publicly support outcomes. RETI assesses and selects CREZ locations based on cost and environmental impact, and it prepares detailed transmission plans for each zone; regulatory authorities issue final licensing decisions. By ensuring that the Stakeholder Steering Committee reflects a diverse range of interests, the active support for RETI decisions from all members provides credibility to its proposals, minimizes opposition, and facilities decision-making by regulatory authorities.[5] Other best practices for public engagement include:

1. Involve from the outset of planning public stakeholders that reflect many perspectives; engage them throughout the process
2. Use a transparent process for developing routing options[6]
3. Explain the objectives for grid expansion, especially as it relates to public concerns (e.g., reliability, electricity prices, RE goals, employment)
4. Present alternative solutions and their trade-offs
5. Clearly describe the types and distribution of costs and benefits, as well as costs of inaction or suboptimal actions (REALISEGRID 2011)
6. Create a publically approved, transparent process for evaluating property values and compensation levels
7. Create a regulatory approach that is accessible to the public and minimizes burdens on applicants
8. Require support from national political bodies for international projects to proceed; authorization for such projects could follow a simplified process, possibly at only the national level (REALISEGRID 2011)

Text Boxes 2-4 provide examples from the case studies of additional approaches to engaging the public and facilitating new transmission.

**Text Box 4. Example of Approach to Public Engagement—
Denmark**

To address public concerns about aesthetics and impacts on land uses, Denmark intends to bury its entire high-voltage grid.

A Cable Action Plan prepared by Energinet and its regional partners analyzed the undergrounding of the entire 132 kV– 150 kV transmission grid and all *new* 400 kV lines by 2030. The committee also suggested that it is probable that the entire distribution network (6 – 60 kV) would be buried in this timeframe. (Danish Electricity Infrastructure Committee 2008)

The cost of undergrounding the 132 kV – 150 kV transmission grid was estimated to be $2.0 billion, while the cost of burying the distribution grid was estimated to range from $1.7 billion to $2.0 billion. (Danish Electricity Infrastructure Committee 2008).

Challenges to Implementing Best Practices
Integrating the meeting-intensive stakeholder process that drives decisions at the local level with a streamlined and expedited process for approving large transmission projects—especially international projects—represents a challenge.

Actions to Improve Public Support
1. Lead public engagement and communicate with the public about why new transmission is essential.
2. Encourage the adoption of approaches that can facilitate new transmission builds

B. Coordinate and Integrate Planning

Rationale
Planning is a critical element of all power systems, but this is especially the case with variable RE, which requires a flexible system—one that can respond to expected and unexpected changes in demand and supply. Coordinated and integrated planning helps decision makers anticipate how

variable RE might impact the grid and its operations, and what options would optimize costs across a system. Planning that is segregated by type (generation, transmission, and system performance) or geography adversely impacts the ability to employ best practices to accommodate RE, including diversifying RE locations, enlarging balancing authorities, and increasing system flexibility. Plans for ensuring physical capacity are also often considered independent of alternative institutional or market structures. These alternatives, however, could significantly increase access to existing and planned capacity, and they could lower the overall costs of accommodating variable RE.

Text Box 5. Example of Market-based Approach to Planning — Australia

Australia uses market-based cost differentials to guide generation and transmission development, and draws from a national, rather than regional, examination of network development. The Australian Energy Market Operator (AEMO) develops a National Transmission Network Development Plan (NTNDP) to inform developers of the most economically efficient locations or configurations for new projects, but there are no requirements to build in the areas identified by the NTNDP. The NTNDP incorporates benchmarked and accepted new technology and network development costs. While the generation development path is only indicative, it highlights regions where development is expected to be economically efficient. Market-based prices provide further signals on economic efficiency. Spot market prices include a location-specific multiplier on the regional price that is applied to all connection points within each region to reflect losses. Connection costs and congestion-based pricing through the spot markets provide additional price signals to new generation. One challenge to this approach, however, is that developers may not always choose the most economically efficient locations or configurations. (See Appendix B for details.)

Best Practices

Planning comprises an inherently complex set of activities that are undertaken by multiple groups and jurisdictions for a given power system. The approach to planning differs widely— Australia (see Text Box 5) uses market-based cost differentials to guide development while Texas (see Text Box 6) draws on its clear jurisdictional authority to centrally plan. Each jurisdiction, however, can benefit from a few key principles that help accommodate variable RE:

1. Integrate the planning of generation, transmission, and system performance
2. Ensure institutions and markets are designed to enable access to physical capacity
3. Build from local and regional planning to better integrate and coordinate information across jurisdictions

Integrate the Planning of Generation, Transmission, and System Performance

Planning that optimizes resources across an entire network—generation, transmission, and system operation—better anticipates and minimizes system flexibility needs to accommodate variable RE. For example, an aspect of system performance—line rating—can be made more accurate through sensors or dynamic line ratings, which account for the positive correlation between wind generation and greater transmission capacity[7] due to cooler lines, and can offer a potential capacity increase of up to 50% (IPCC 2011). A study of ERCOT's use of dynamic ratings in Texas concludes that the ratings can reduce congestion constraints and the number of uneconomical generators necessary to alleviate congestions, and thereby lower overall rates (Hur et al. 2010).

Text Box 6. Example of Centralized Approach to Planning—Texas

In Texas, the need for new transmission was identified after heavy wind development occurred in an area with weak transmission, resulting in frequent curtailments. To increase capacity and accommodate revised renewable portfolio standard goals, in 2005, the Texas legislature passed a bill initiating the Competitive Renewable Energy Zone (CREZ) process, which made two key changes: 1) the Public Utilities Commission of Texas did not have to prove need, through financial commitments by generators, to certify new lines; 2) transmission developers do not absorb cost of lines if capacity is underused. Nevertheless, developing just-in-time transmission can be challenging. Wind development in Texas did not occur as planned due to many factors, including decreases in natural gas prices and an expected expiration of the federal production tax credit.

Significant to increasing RE penetrations is to align the planning timescales for generation and transmission to allow the two to develop in coordination. For example, renewable energy zones can be used to try to plan transmission to areas where there are high quality renewable resources (see Text Box 6), alleviating the need to construct generation prior to the more time-consuming transmission.

Finally, planning, which has historically been based on snapshot analyses of periods of greatest system risk, could better account for variable RE by looking at how risk varies at different times in the year and accounting for correlations between RE resource patterns and daily and seasonal loads (IPCC 2011, building on analysis by Burke and O'Malley 2010).

Ensure Institutions and Markets Are Designed to Enable Access to Physical Capacity

Planning often focuses on physical capacity; yet changes to market and institutional structures may more readily access resources needed for system flexibility. For example, hourly dispatch pushes ramping to meet the next hour's schedule into a 20-minute period. This in turn reduces access to physical flexibility and increases the need for regulation. Shifting to sub-hourly dispatch can improve the efficiency of operations and reduce the need for regulating resources (Milligan and Kirby 2010a).

Similarly, physical transmission rights can prevent the efficient use of a transmission asset by allowing an entity to own physical rights that are then inaccessible to other generators. Converting physical to financial transmission rights allows a market-based mechanism to access this capacity during congestion, reducing the need for new transmission.

Build from Local and Regional Planning to Better Integrate and Coordinate Information across Jurisdictions

A broad network expands transmission available for capacity and reliability needs, and reduces the need for reserves. Two challenges make integrated planning across a region difficult— defining the planning objective and allocating costs from improved reliability. As an example of the first challenge, least cost system planning over a broad region could suggest new interconnections that in turn raise generation prices in historically low-price areas (as in Colorado), creating a conflict among jurisdictions about whether least cost planning should be the objective. As an example of the second challenge, an integrated network with improved interconnections increases reliability, but many power systems lack a universally accepted way to allocate costs from infrastructure improvements that are located in one region but improve reliability for an entire network.

The case studies illustrate the significance of legal and institutional structures in formulating approaches to regional integration. Ireland's transmission system is state-owned; this has facilitated the shift in planning from local to national levels. Texas was able to make a legislative change authorizing the Public Utilities Commission of Texas to establish renewable energy zones and approve transmission lines to serve these areas, in advance of generation commitments. In contrast, the western United States, which lacks a regional transmission organization (RTO), must navigate regional planning through the Federal Energy Regulatory Commission, state planning commissions, and the Western Electricity Coordinating Council (WECC).[8] Europe has made efforts to overcome their jurisdictional overlap by creating the European Network of Transmission System Operators for Electricity (ENTSO-E), which is obliged to prepare ten-year network development plans (non-binding) that include all of Europe. Also, a coalition of grid operators and environmental groups has established the RenewablesGrid-Initiative, which seeks a 100% RE grid by building public support and establishing pilot projects.[9]

The U.S. Department of Energy has supported a coordinating process by providing regional grants to develop long-term generation and expansion models that reflect increased complexity and scope. WECC, ERCOT, and the Eastern Interconnection were able to analyze impacts on cost, dispatch, ramping, transmission use, operational reliability, for example, across each of their interconnections from changes to policies, planning horizons, and grid improvements. As part of this support, U.S. DOE also provided money to state regulatory bodies within the interconnections to interface with the studies.[10] An important early step towards interconnection-wide planning is to coordinate base case models across utilities, RTOs, and independent system operators (ISOs), including information from neighboring systems.[11] Text Box 7 provides an overview of grid integration studies.

Challenges to Implementing Best Practices
- Coordinating multiple jurisdictions, especially when planning objectives and impacts of integration differ across a region
- Ensuring that each institution has the capacity to communicate with and integrate planning materials from multiple jurisdictions
- Reconciling the planning for a specific generation target (e.g., 20% wind) while also planning for transmission that reflects a range of

possible outcomes (e.g., 30% wind with its added flexibility requirements)
- Translating analyses[12] into action, particularly when legal framework prevents change

---

Text Box 7. Grid Integration Studies

- Europe: Towards a Successful Integration of Large Scale Wind Power into European Electricity Grids. (EWIS 2010).
- Germany: Grid Study II – Integration of Renewable Energy Sources in the German Power Supply System from 2015 – 2020 with an Outlook to 2025. (Dena 2011).
- Ireland: All Island Grid Study. (DCENR and DETI 2008).
- Ireland: TSO Facilitation of Renewables Studies. EirGrid and SONI (2009).
- United States: Western Wind and Solar Integration Study. (National Renewable Energy Laboratory 2010).
- United States: Eastern Wind Integration and Transmission Study. (National Renewable Energy Laboratory 2010).
- For additional integration studies, see the Utility Variable Generation Integration Group's wind integration library: http://www.uwig.org/opimpactsdocs.html

---

Actions to Improve Planning Coordination

1. Share best practices and guidelines for adapting advanced planning capabilities to accommodate high penetrations of variable RE
   - Enhance capacity of institutions to increase integration, complexity, and coordination of—and stakeholder participation in—planning
   - Provide vision and empower leadership to realize how to move from analyses and recommendations to actions
2. Convey to all stakeholders the need to review existing rules and methods for planning, design, and operation to accommodate higher penetrations of variable RE; including:
   - Utilities, RTOs, and others: Adapt or develop tools that enable system planning to incorporate requirements for flexibility and present recommendations[13]
   - Regulatory commissions: Use rewards and punishments (e.g., rate recovery, rules) for coordination

C. Develop Rules for Market Evolution that Enable System Flexibility

Rationale

Markets help minimize power system costs, and for systems using variable RE in particular, they can facilitate access to a range of options that increase system flexibility. Higher penetrations of variable RE require increased flexibility from the power system to manage the variability and uncertainty of the generation. Flexibility can be achieved through changes in market operations, increased transmission, or the addition of flexible resources to the system, such as more flexible generating units, storage, and demand response. While flexibility is of high value to the system and can reduce the need for new capacity, it may come at a cost to power suppliers. Increased ramping of units that are not adequately designed for cycling can result in maintenance issues or reduce the lifetime of units. Also, conventional generators may experience profit margins that are insufficient to maintain their long-term financial viability if variable generators depress wholesale market prices and generators are only compensated for energy production. Therefore, market rules and operations may need to be modified over time to achieve operational efficiency in systems with increasing penetrations of variable RE.

Text Box 8. Example of Approach to Market Design—Australia

Australia's National Electricity Market (NEM) covers 92% of Australian electricity demand and uses a complex spot market pricing signal to reward contributions to flexibility— beyond what can be offered through an energy-only price. The NEM pricing signal values diversity, alignment of generation with load profiles, and location (e.g., distance from load; reductions in congestion).

NEM has been operating since 1998, before wind farms came online in Australia. Key principles of market operation, which have been consistent over time, help accommodate variable renewable energy, though they were not explicitly designed to do so. NEM is a fast market; it is a self-commitment market with 5-minute dispatch. Offers can be changed any time prior to dispatch, and there is a 5-minute regional clearing price and 30-minute settlement period. NEM also allows for a negative price of -$A 1,000 per MWh, which provides an incentive against oversupply.

NEM has been amended over time to incorporate variable generation. The rules of NEM were amended on the basis of evidence-based analysis, in consultation with industry and with support of, but independent from, State and Australian Governments. Previously, two dispatch categories ("scheduled" and "non-scheduled") existed; wind was categorized as "non-scheduled." A new category, "semi-scheduled" was created to incorporate variable generation. Semi-scheduled generators submit offers for their output and are included in constraints and market dispatch. If a wind farm is part of a binding network constraint, its output can be capped.

All market participants receive forecasts of total regional wind contribution, customer demand projections, and potential network constraints, enabling them to make output and commitment decisions to suit. (See Appendix B for details.)

Best Practices

Use markets to support the most cost-effective solution to increasing flexibility, which could include:

Flexible Generation

Encourage sub-hourly scheduling and dispatch intervals (5- or 15- minute) and shorter gate closure periods to improve system efficiency. Markets using hourly scheduling or areas without organized markets that rely on bilateral contracts with fixed hour energy delivery can be problematic with higher penetrations of RE. Fast, sub-hourly markets that operate on 5-minute or 15-minute intervals, such as in Australia (see Text Box 8), are preferred for integrating variable generation because they can more effectively address changes in the variable generation and the load ("net load"[14]) (EWIS 2010). By dispatching the system in shorter increments, the movements of conventional generators needed to balance the system (i.e., regulation) can be reduced. Also, the system can run more efficiently and minimize reserve requirements.

Text Box 9. Example of Approach to Market Design — Denmark

In Denmark, market design encourages system flexibility through expanded power markets, fast market design, CHP, and negative prices.

- West Denmark joined the Nord Pool power market (which includes Norway, Sweden, and Finland) in 1999, followed by East Denmark in 2000. Electricity traded through Nord Pool has increased from 45% in 2006 to approximately 75% (VTT 2007). In 2010, power markets were further connected with the establishment of the Interim Tight Volume Market Coupling (ITVC), which connects Nord Pool and the Central West European market.

- Through faster market design, two markets, Elbas (an intraday market, with continuous trading up to 60 minutes ahead of delivery) and the Regulating Power Market (operating up to 15 minutes before delivery), provide opportunities for producers, retailers, and consumers to provide short-term flexibility and remedy imbalance in trading positions.

- Combined heat and power (CHP) plants, representing more than 55% of all electricity production (Danish Energy Agency 2011), support flexibility through direct participation in the power markets and as a source for flexible storage. CHP plants are required to participate in the spot power market, and around one-third of small CHP plants are active in the Regulating Power Market. When wind is high, CHP plants rely on their storage to continue to provide district heating, without needing to operate and compete with wind generation. When electricity prices are very low or negative, it is now cost effective to use cheap (wind) electricity to produce the low-temperature steam needed for district heating (Kiviluaoma and Meibom 2010).

- In 2009, negative prices were permitted in the day-ahead market for the first time (previously there had been a price floor at €0/MWh). An alternative to arbitrarily curtailing, negative pricing provides an economically efficient way to reduce output during excess generation. Large consumers paying the spot price will also be encouraged by negative prices to shift consumption to these times.

In addition to moving to sub-hourly dispatch, reducing gate closure times—the point at which the market commits to deliver electricity—minimizes forecast errors and allows for trading at potentially lower cost than power that would otherwise be required for system flexibility. Germany and Denmark have reduced their gate closure times, e.g., Denmark to 15–60 minutes ahead of dispatch (see Text Box 9).

To increase system reliability, use capacity markets to help address concerns about declining wholesale electricity prices. With higher penetrations of variable RE generation, the market clearing price is lowered, as wind and solar will be dispatched first as the lowest marginal cost generators. The marginal units on the system, such as combustion turbine peaking plants, will experience reduced numbers of operating hours and lower compensation, which could make the units become uneconomic over time. However, these peaking units may be highly desirable for increasing system flexibility. Therefore, capacity payments or other financial incentives may be required to more fully compensate these generators for the benefits they provide to the system. Capacity markets are one vehicle for compensating needed generators to ensure that a sufficient amount of flexible generating capacity is available to serve the grid in current and future years. Additional services markets, which have yet to be fully implemented, include flexibility markets to reward generators with fast ramping, ramp rate control, and quick-start capabilities. This may incentivize generators to build-in sufficient flexibility needed by the system.

Use zonal or nodal pricing to help manage congestion on the system and encourage development of resources where needed. Zonal or nodal pricing can send signals to the market about congestion and the need for resources in particular locations. For example, Australia's use of nodal pricing and negative prices for hours when too much energy is on the system has created incentives to diversify the location of renewable generators. Because of negative prices in South Australia, where the wind is not well correlated with system peaks, wind developers are beginning to look to other locations to site projects. The use of nodal pricing is encouraging investment in regions with the highest prices over time (see Appendix B). The use of nodal pricing is also being considered in Germany to help address disconnects between the areas of renewable energy development and loads (see Appendix D).

Develop equitable rules for curtailment of variable generators during periods of excess renewable generation on the system. Denmark has adopted a number of policies to deal with excess wind generation at night when loads are at their lowest levels. The Nordic power market allows negative pricing to occur on the system – meaning that a generator would have to pay to generate at times when the system does not need power. This dissuades wind generators from bidding in too much power at night when the system cannot handle it.[15] Denmark has also established a policy precluding the curtailment of wind generators unless thermal units are being operated at minimum capacity. Land-based wind is compensated for curtailment; offshore wind farms, which may need to be curtailed for reliability reasons in the future, may not be compensated if notice is given the day ahead that curtailment will be needed. Offshore wind turbines are also subject to other types of controls, such as limits on ramps (see Appendix C). Australia and Germany, as described in Appendix B and D, have also implemented negative pricing.

Design imbalance payment rules so that they do not unduly penalize variable generators. Because wind plants cannot predict their output—particularly in the day ahead market—utilities or market operators, for example, may subject generators to imbalance penalties for deviating from their generation schedules. These penalties are typically designed to prevent market manipulation in organized markets, but in some cases may be unduly stringent for variable generation sources, making them uneconomic. In the Nordic market, wind generators only pay extra for imbalances during hours in which their deviations are in the same direction as the net system imbalance (see Appendix C). The extra cost assessed reflects the cost of generation used to balance the system during that hour. Additional fees are not levied on wind generators if their deviations are beneficial to the power system.

Consider hybrid market solutions to increase flexibility in areas without organized markets. For example, the Energy Imbalance Market, under consideration in the western United States, where there is no central organized wholesale market, is a partial market-based solution that would help the region manage higher penetrations of variable renewable generation. The proposed Energy Imbalance Market would automate and speed up the process for transacting imbalance energy. It would also enable more efficient power system operations by enabling balancing areas to access the most cost effective resources in the region to balance the system and manage transmission congestion (see Text Box 14 and Appendix G).

Require flexibility in resource planning or provide financial incentives to ensure new capacity is as flexible as possible. In an integrated planning market, utilities can consider flexibility criteria when evaluating bids for new generation resources. This would ensure that generation added to the system would be as flexible as possible and enable the system to more easily accommodate higher penetrations of variable RE. Another way to promote flexible resources is to provide direct financial incentives to encourage resources to be more flexible. Germany has developed a fund to encourage new fossil-fired power plants to use the most flexible technology available to maximize their ability to ramp to meet the system's balancing needs (see Appendix D).

Flexible Storage

Storage, such as pumped hydro, batteries, flywheels, and compressed air, contribute to system flexibility in three ways (NERC 2010):

1. Load shifting: Storage can absorb surplus RE power and reduce curtailments, and reduce the net flexibility required of conventional units; strategically placed storage can also reduce transmission congestion (IPCC 2011)
2. Balancing: Over a course of minutes to hours, storage can smooth net load and reduce the need for spinning reserves
3. Frequency and regulation: The cost-effectiveness of storage can be improved by allowing fast-discharge storage to participate in ancillary markets, such as regulation, demand following (ramping), and capacity (NERC 2010, Delille et al. 2010).

Ensuring the optimal use of storage, e.g., supporting the power system as a whole rather than dedicated to a single generator, further improves its cost-effectiveness (NERC 2010).

Resources that serve non-electric demand may be a particularly cost-effective and large source of storage and system flexibility (Kiviluaoma and Meibom 2010). Electric heat boilers, operating during periods of low to negative prices (due to very high winds), can store heat and reduce curtailments of wind generation. Combined heat and power (CHP) plants, which also offer heat storage but primarily from burning fossil fuels, are more likely to shut down during high winds when fossil fuel costs are uncompetitive. Shutting down allows more room for wind generation; yet CHP plants can continue to serve heat demand through thermal storage (Kiviluaoma and Meibom 2010).

Denmark uses the heat storage of CHP plants, representing more than 55% of total electricity production, to complement variable RE (see Text Box 9). CHP plants produce low-temperature steam for district heating, which allows for easier and more considerable storage than the high temperatures needed for industrial processes. When wind generation is high, storage is essential for enabling CHP plants to shut down and allow higher penetrations of wind generation. To further contribute to system flexibility, CHP plants are required to participate in the spot power market, and a third of small CHP plants participate in Denmark's ancillary markets. Germany (see Text Box 10) is also considering incentives for CHP.

Text Box 10. Example of Approach to Flexible Storage—
Germany

Germany has implemented mechanisms to encourage energy storage. There is a €200 ($261) million budget for storage R&D up to 2014, and new storage facilities are exempt from grid charges and the EEG levy (BMU 2011).

Germany has demonstration plants in operation that examine the benefits of using excess electricity production to produce hydrogen, and hence synthetic methane gas, which could be fed into (and so stored) in the gas grid (Netzentwicklungsplan Gas 2012).

Germany is also considering mechanisms to encourage CHP. One option under consideration would provide an incentive to CHP plant owners for including additional thermal storage. A second proposal would provide grants covering 30% of the costs for additional heat storage, with the goal of increasing CHP power production from around 15% in 2010 to 25% in 2020 (Argus 2011).

Flexibility of Load through Demand Response and Use of Smart Grid Technology

Demand response, also referred to as load management, contributes to system flexibility in the same three ways as storage:

- Load shifting: Flexible loads can participate in capacity markets, and reduce the flexibility needed by conventional units. Loads, especially through incentives such as low and negative pricing, can shift to times of surplus RE power, thereby reducing curtailments.
- Balancing: Load has the potential to quickly offset large losses of power, as occurred in Texas in February 2008, when demand response reduced load by 1,108 MW in 10 minutes (see Text Box 11). Load such as water desalination, ice production, electric vehicles, and thermal loads (e.g., water heating) are well-suited to this (IPCC 2011, Milligan and Kirby 2010b, Kirby 2007).
- Frequency and regulation: Load can participate in ancillary markets. Demand response for frequency regulation has been important for a small, isolated system like Texas. For example, as part of demand response in Texas, load moves up and down automatically (within 1/3 second) to maintain frequency at 60 Hz; participates in nonstandard reserves by being able to ramp load in 30 minutes; and, as "spinning"

(responsive) reserves, must be able to respond within 10 minutes of a hotline call. ERCOT procures in the ancillary market 2,300 MW each hour, representing the equivalent of the two largest units (nuclear), in case they trip off-line. ERCOT caps load's participation in this market at half (1,150 MW). Participating loads, which generally bid at $0, allow generators to set the clearing price, with an average in 2010 at $20/MWh, reflecting low natural gas prices (Wattles 2011).

Both price signals and grid management rules can enable power systems to access load as a source of flexibility (NERC 2010). To be effective, adequate communication infrastructure between system operators and load is necessary (IPCC 2011).

Text Box 11. Example of Approach to Flexible Demand—Texas

Demand response participation provides ERCOT with flexibility at a low cost. After the introduction of retail competition, Texas wanted a mechanism to continue to draw on the interruptible load that had been participating in special tariffs. The restructured market allows demand response to participate in ancillary services markets. In February 2008, when anticipated wind and traditional generation fell short, and demand ramped up more quickly than anticipated, 1,108 MW of demand response were activated in 10 minutes (ERCOT 2008). Since 2006, demand response has been deployed 21 times in Texas (Wattles 2011).

Challenges to Implementing Best Practices

- Physical solutions to increase flexibility are straightforward; institutional, legislative, and market barriers will be the challenge
- Adjusting market rules is a time-consuming and stakeholder-intensive process. Various stakeholders have different economic interests, but it is important to develop solutions that lead to greater system efficiency overall while allowing regulatory flexibility to distribute gains equitably
- Implementing solutions for areas without organized wholesale electricity markets can be costly, and obtaining consensus among diverse stakeholders to implement partial market-based solutions may be challenging.

Actions to Use Markets to Enable System Flexibility
1. Lead development and innovation of market design options for enabling higher penetrations of variable RE generation (e.g., commission studies to identify potential impacts of variable generation on electricity markets and generator compensation and identify needs)
2. Encourage market operators to adopt rules to improve system efficiency with higher penetrations of variable RE. Work with regulators to educate stakeholders about best practices
3. Play a leading role in negotiating a framework for integration that optimizes flexibility across regions
4. Partner with the private sector to advance development and demonstration of technologies and tools that increase flexibility (e.g., fast ramping, storage, smart demand response), all in coordination with market design innovation

D. Expand Access to Diverse Resources and Geographic Footprint of Operations

Rationale

One of the concerns about integrating variable RE is vulnerability of the power system to weather events. Integration studies have consistently found that expanding access to diverse resources reduces this vulnerability. This can be achieved in two ways: enlarging effective balancing areas and diversifying the location and types of RE generation.

By enlarging balancing areas, the relative variability and uncertainty in both the load and RE generation will be lowered, smoothing out differences among individual loads and generators. This in turn reduces the need for reserves and lowers overall integration costs. Larger balancing areas may also provide access to a greater amount of flexible generation.

Greater geographic distribution of renewable resources reduces the variability of RE because weather patterns are less correlated across large geographies, reducing the relative magnitude of output changes. Greater diversity of technologies similarly reduces the correlation among generators, and thus has an effect that is similar to that of increasing geographic diversity.

Best Practices

Creating larger balancing areas can help integrate higher penetrations of variable RE generation on the system. Larger geographic areas also tend to reduce aggregate forecasting errors (NERC 2011). Denmark, which has already achieved relatively high levels of wind energy penetration with 22% of its production from wind power in 2011, has benefitted from participating in the larger Nordic power market to manage the variability. In this way, Denmark can access flexible hydropower and other resources to provide reserves to accommodate the variability of wind. The pooled operation minimizes the need to curtail wind and reduces reserves required by the system (Holttinen et al. 2009). Denmark also has significant efforts underway to expand its electrical linkages to other nearby markets, including to Norway and the Netherlands. This is expected to help decrease power prices in other areas and enable Danish wind generation to access pumped storage hydropower resources (see Text Box 12).

Text Box 12. Example of Approach to Diversification—Denmark

Denmark has enlarged its balancing area through electrically linking to adjacent and nearby markets. East Denmark is part of the Nordic system, and while both East and West Denmark have long interacted through the Nordic power market, in 2010, they were electrically connected through a DC link, providing additional management options.

Electrically linking Denmark's markets helps spread wind power production more widely, increasing its value, and enabling surpluses to be stored – for example, in Norwegian pumped hydropower facilities.

Interconnecting isolated, small systems with neighbors allows access to generation sources from larger grids. Systems can more effectively balance higher concentrations of variable generation if they can access additional balancing resources in other areas. For example, Ireland is a relatively small island system with few regional interconnections. However, Ireland has been taking steps to increase its interconnections to be able to handle a large amount of wind on its system. The East-West connector, which is a 500 MW high voltage DC link to Wales, is currently under construction and is expected to decrease wind curtailment levels. Ireland has also taken steps to expand its electricity market, by creating a Single Electricity Market with Northern Ireland in 2007, despite the use of different currencies. This expanded market has created a larger pool of resources for balancing the system and has provided Ireland with more transmission connections to the United Kingdom (see Text Box 13).

Text Box 13. Example of Approach to Enlargement of Balancing Areas—Ireland

Ireland has expanded its balancing area through the creation of a Single Electricity Market (SEM) with Northern Ireland in 2007. All electricity above 10 MW sold and bought in Ireland is traded through the central electricity pool of SEM through a market clearing mechanism. The mechanism is unique in that it is operating with dual currencies and jurisdictions. It was established through a series of agreements between regulators and transmission system operators in both countries. Ireland is also developing an interconnector with the UK, which is scheduled to be complete in 2012. An auction management platform that will allow participants to trade across the interconnector is being developed.

For areas without organized markets and with small balancing areas, using hybrid market solutions can achieve balancing area cooperation and reserve sharing. These cooperative mechanisms can result in cost savings from sharing reserves without the need to create a fully organized market structure. Studies of the proposed Energy Imbalance Market (EIM, described in section C) in the western United States find that it would reduce the reserve requirements substantially if widely adopted in the region, resulting in significant cost savings (King et al. 2011, WECC 2011). By expanding the balancing area, the EIM would also reduce the variability of the renewable generation by enabling access to renewable energy plants in different geographic locations. By managing imbalances over a broader region, the EIM can net out differences between generation and load across a broader area (see Text Box 14).

Because transmission access often influences where RE generators are located, renewable energy zone planning can help identify diverse areas of RE resources and encourage transmission planning to those resources. For example, renewable energy zones have been used in Texas and the western United States to facilitate the development of new transmission projects to serve areas with the greatest potential for developing renewables. In Texas, in particular, the use of renewable energy zones helped identify needed transmission paths, align the transmission and generation planning timelines, and streamline the process for approving transmission projects (see Text Box

6). The zone efforts have been effective in rapidly bringing on-line new transmission projects and diversifying the development of wind to various areas of the state from an initial concentration of development in West Texas.

Text Box 14. Example of Approach to Enlargement of Geographic Footprint of Operations—U.S. West

In the western United States, there is not an organized wholesale electric market. In order to facilitate trading without creating a full wholesale electricity market, the Energy Imbalance Market has been proposed. The EIM would allow balancing areas to cooperate, thus resulting in operational cost savings, because fewer reserves would be needed to manage the system. The Western Electricity Coordinating Council helped develop a high-level design specification and evaluation of the EIM proposal, but the EIM will need support from state public utilities commissions in order for utilities to be confident that they will get cost recovery for participation in the EIM. Technical studies have been partially funded by the Department of Energy.

Incentives may be provided to encourage RE to be sited in diverse locations if this minimizes total system costs (i.e., the total cost transmission plus generation). Policies can encourage geographic distribution of resources, but this is only worthwhile when the increased cost of diversifying to poorer resource areas is less than the savings associated with reduced costs for transmission upgrades to better resource areas.

Project bid evaluations could include an assessment of the location of the resource and its potential impact on the system, thereby encouraging a mix of resources on the system. In markets where integrated resource planning is used, the location of the renewable energy and its potential impact on the system could be examined in the bid evaluation process. This could ensure that not all variable generation is located in the same region where it is affected by the same weather events.

Challenges to Implementing Best Practices
- There are significant institutional challenges to achieving balancing area cooperation or consolidation. Many stakeholders with different objectives and concerns about consolidation may be involved.
- Another challenge for areas with multiple balancing areas and without organized markets is shared telemetry. There is a need to determine

what information is shared, what is automated, and how to address the cost.

- Diversifying locations may necessitate operating in regions that are not as cost-effective (i.e., projects in less windy areas).

Actions to Expand Access to Diverse Resources

1. Support study and evaluation of methods to increase balancing area size or balancing area cooperation, particularly for areas with small or disaggregated balancing areas
2. Convene stakeholder discussions to evaluate options and identify needs for overcoming institutional challenges in merging or increasing cooperation among balancing areas
3. Support cooperation among TSOs[16]
4. Lead the renewable energy zone planning process for transmission to different resource zones; this could be done to encourage diversity
5. Ensure resource assessment is state-of-the-art for all RE resources
6. Encourage utilities to consider and evaluate the location of new RE generators and encourage diversity through the bid evaluation process

E. Improve System Operations

Rationale

Beyond market and institutional changes to system operations described in earlier sections (e.g., faster scheduling, enlarged balancing areas), system operations can be improved by adopting advanced forecasting techniques and changes to grid codes. Using advanced forecasting techniques helps reduce the amount of system flexibility needed to integrate variable RE generation. Renewable energy generation can be variable, changing with the time of day and weather patterns, and uncertain because of the inability to predict the weather with perfect accuracy. Using forecasts in grid operations can help predict the amount of wind energy available and reduce the uncertainty in the amount of generation that will be available to the system.

Revising grid codes to address issues related to variable generation (e.g., concerns about frequency control and other disruptions to network stability) both allows hardware and procurement agreements to be designed in advance to support the power system and reduces the financial burdens of retroactive requirements. Creating a model grid code can serve as a guide for each system to evaluate what changes are needed.

Best Practices

Advanced Forecasting

Integrate forecasts into fast market operations, the control room, and other standard operating practices of the system operator or market operator. Forecasts are more accurate the closer they are to real time. System operators use accurate forecasts to determine unit commitment and reserve requirements; this can minimize ramping requirements of fossil plants and the need for reserves—a cost savings. For example, the Western Wind and Solar Integration Study found significant savings with the use of advanced forecasting techniques in the western United States (NREL 2010). Denmark integrates forecasts into system planning up to 2 hours ahead of the time of operation, but is moving toward 5 minutes (see Text Box 15).

Text Box 15. Example of Approach to Improvement of System Operations—Denmark

Energinet has greatly enhanced its Energy Management System (EMS) to monitor real-time performance of the power system. This includes real-time estimates of wind power to be fed into the grid. This is an important aspect as most wind power in Denmark is connected at the distribution level, which is passively managed, so it is not very visible to the system operator.

Energinet, Denmark's system operator, uses multiple and advanced forecasts. Forecasts are used in planning of system operation, day-ahead (cross-border) congestion management, the commitment and economic dispatch of controllable power plants, contingency analysis, the assessment of grid transfer capacity, and the need for regulating power.

Energinet's Drift *Planlaegnings* (Operational Planning) System (DPS) tool integrates forecasts of wind and CHP output into system planning up to two hours ahead of operation—and in the near future up to five minutes. The DPS also provides the operator with a view of power flows between the Danish system and those of Germany, Norway, and Sweden. All power plants greater than 10 MW must provide five-minute updates of power production. Wind generators receive financial compensation for producing what is forecast and sold in the day-ahead market.

Grid codes were revised in 1999 to require wind turbines on the high voltage grid (mainly off-shore wind) to remain connected to the grid so that they can support the grid in case of fault.

> Given that 90% of wind turbines are connected at the medium voltage (60 kV) level and below, similar grid codes now apply at that level also. New regulations allow off-shore wind farms to be controlled, if necessary, by placing a ceiling on output, reducing output over a fixed period, smoothing the rate at which output increases, or reducing output by a fixed amount over time.

Ensure RE plants continually provide updated data on power, wind speed, and turbine availability to system operators to improve the accuracy of the forecasts they use. Frequently updated and comprehensive data are needed for system operators to use forecasts in real time. One important data issue is that system operators need to be aware whether turbines are offline for maintenance; a lack of turbine availability data significantly impacts the quality and usefulness of the forecast data. Comprehensive data on power output, wind speed, and turbine availability are needed on a real time basis. Also, the data used for forecasting wind plants needs to be able to adequately capture differences among individual turbines at the plant, particularly for large plants with mixed terrain. A small wind farm may be adequately captured by data from a single tower, while large projects may need multiple sources of wind speed data (Grant et al. 2009). The ability for wind farms to reliably transmit this information frequently to grid operators is important. For example, Australia requires all transmission-level wind farms to provide a set of forecasts from five minutes to two years (see Text Box 16).

> **Text Box 16. Example of Approach to Advanced Forecasting—Australia**
>
> Early in the development of wind power in South Australia, the importance of forecasting was recognized. However, a number of forecasting proponents were significantly concerned about funding and continuously operating individual forecasting systems. The South Australian Electricity Supply Industry Planning Council and the market operator initiated working groups and engaged the Australian Government as a potential source of funds. As a result, a forecasting model was developed that incorporates inputs from, e.g., individual wind farm or turbine outputs, on-ground automatic weather stations, and satellite imagery. In order to drive efficient outcomes, the market operators use the wind forecast, along with a future demand forecast and advanced generator bids to provide a forward price projection up to seven days ahead.

Generators use the price forecasts to decide when to commit their plant. Near-term plans for improvement focus on developing large ramp forecasts. (See Appendix B for details.)

Using multiple forecasts can be beneficial for system operations. The use of balancing area and project-level forecasts can improve accuracy and encourage better decision making by generators (in bidding) and system operators. Having project-level and balancing area forecasts improves how transmission flows are managed on the system. In addition, many forecasting systems use a blend of several forecasts instead of a single forecast to help address uncertainty and to improve accuracy. State-of-the-art forecasts today generally use a combination of statistical and physics-based atmospheric models. Short-term forecasts (up to 6 hours) often use statistical methods based on recent data from the wind plant or locations nearby. Longer-term forecasts generally rely more on numerical weather prediction models, which forecast weather based on physical properties of the atmosphere. Combining methods can reduce forecast errors. Also, the level of accuracy increases when combining forecasts for larger areas (Holttinen et al. 2009).

Continued evaluation and improvement of forecasting methods can facilitate more efficient operations and help address higher penetrations of variable generation. One area of improvement has been the development of ramp forecasts, which can be important for managing system operations. Because wind ramps can be affected by a variety of weather events, it can be a challenge to provide meaningful information to grid operators, but advancements are being made and more systems are incorporating ramp forecasts (Holttinen et al. 2009). For example, Australia is improving its centralized forecasting program by developing large ramp forecast capabilities (see Text Box 16). Also, ERCOT, the grid operator for Texas, recently implemented a ramp alert system to help prepare for large and sudden changes in wind energy output.

Grid Codes

As wind and other variable sources increase in penetration, their ability to support the grid during moments of grid instability becomes increasingly important. Grid codes, which are rules established by the system operator to govern how generators connect to grid, help ensure that variable RE is compatible with, and can help contribute to the stability of, the power grid (IEA 2011). A necessary first step is to evaluate existing rules to determine if new approaches to planning, design, and operation are needed for high RE.

As an example of this first step, the NERC created a roadmap of system reliability requirements based on an integrated review of needs and capabilities. The Integration of Variable Generation Task Force, which created this roadmap, had goals of: 1) understanding the operational and planning landscape of systems with variable RE; 2) identifying new approaches to planning, design and operations; and 3) mapping these approaches against existing NERC's reliability standards to identify gaps (Lauby et al. 2011). The task force produced its summary report, *Accommodating High Levels of Variable Generation* in 2009, and suggested work plans for ongoing investigation. The working groups then created recommendations for NERC to consider.

Some of the technical recommendations for grid codes that emerged from this and other studies include:

- Require fault ride-through capabilities. For example 97.5% of wind farms in Spain are able to continue to contribute electricity to the power system during faults (Holttinen et al. 2011). Germany and the US, the latter through FERC Order 661-A (2005), also require fault-ride through capabilities (IPCC 2011).
- Require turbines to provide reactive power, and, in some cases, voltage and frequency control (see Text Box 17).

Text Box 17. Example of Approach to Improvement of System Operations—Spain

Spain has created several procedures to address some of the technical challenges to grid security in accommodating RE on an isolated grid. The innovative Control Centre for Renewable Energies (CECRE) monitors RE installations on the basis of real-time information availability. RE installations must provide real-time telemetry each 12 seconds as well as voltage control following orders of the TSO. CECRE uses a Maximum Admissible Wind Generation tool to determine whether the present generation scenario is acceptable for system operations.

The TSO is responsible for granting access and connection permits for systems connecting to the high-voltage network; the TSO cannot prohibit access for reasons other than the system's operation security, and even in such case, the regulator is in charge of solving the problem.

A procedure for responding to voltage dips was approved in 2006. Wind farms with capacities greater than 10 MW must provide reactive power support; this was extended to solar PV installations with capacities greater than 2 MW in 2010. A power factor range was also established for RE, with penalties for non-compliance and potential bonuses for maintaining a favorable range.

A new operational procedure to maintain optimal voltage control has been proposed. CECRE has made it possible for more than 50% of electricity demand to be met by wind energy over the course of several hours—a particular challenge for a country with an isolated grid)

When evaluating grid code needs, it is important to distinguish between what needs to be addressed at the project level (e.g., wind plants) from what needs to be addressed at the generator level.

To further strengthen system operations, grid operators have also identified several changes that they can implement in the short-term, independently of new regulations and policies, such as visualization of system conditions and automated decision support tools (Jones 2011).

Challenges to Implementing Best Practices
- The use of forecasts requires operational change. Grid operators need to be aware and convinced of the benefits of integrating forecast data in daily operations.
- Optimal grid codes to accommodate variable RE require sophisticated analysis capability and communication infrastructure. Decision-making systems need to be updated, and new programs are required to develop institutional capacity.
- Although codes set in advance minimize retroactive requirements to generators, codes set when penetration is low can unnecessarily burden variable RE technologies. Striking a balance in this timing represents a challenge in providing sufficient notice of new code requirements.
- Using new codes that reflect state-of-the-art engineering-based practices may be better than building from existing codes that draw from older or out-of-date practices.
- Every system is different, which prevents a uniform recommendation and adoption of a model grid codes. System-specific analyses are still needed.

- Jurisdictions have differing standards (e.g., in the United States, NERC requires one standard, WECC might require a higher standard, and a small balancing area can be even stricter). This lack of uniformity challenges manufacturers to serve multiple areas.

Actions to Implement and Improve System Operations
1. Develop national or regional forecasting systems:
- Identify advanced forecasting methods and their benefits
- Provide public support for research and development to improve forecasting methods and put them in the public domain
- Support outreach on forecasting benefits and training on best practices for grid operators
- Encourage efforts to research and continually improve forecasting techniques
- Work with regulators to require that all generators participate in forecasting, which necessitates that generators provide frequently updated data

2. Lead development of codes and standards that meet interregional and international needs to enable greater penetration of variable RE generation:
- Support work with regulatory commissions to evaluate model grid codes, recommend changes, and implement recommendations

SYSTEM-WIDE APPROACH TO AREAS OF INTERVENTION

Areas of intervention are distinct but interrelated; taking a system-wide approach will ensure that not only are individual interventions more effective but also that the system as a whole will be more robust. Figure 2 shows one example of how each area of intervention might relate to others.

COSTS OF INTEGRATING HIGH PENETRATIONS OF VARIABLE RE

Calculating the cost of integrating high penetrations of variable RE is very difficult; however, recent integration studies have demonstrated that the costs are manageable (Bird and Milligan forthcoming). Integration costs can be divided into three categories: those that relate to transmission extension and reinforcement (not including the cost of linking to the grid); those incurred in the balancing of increased volatility in the power system; and those that may be incurred to maintain the adequacy of the power system (i.e., its ability to cover peak demand). Milligan et al. (2011) note that all generation sources, including non-renewable sources, have associated integration costs.

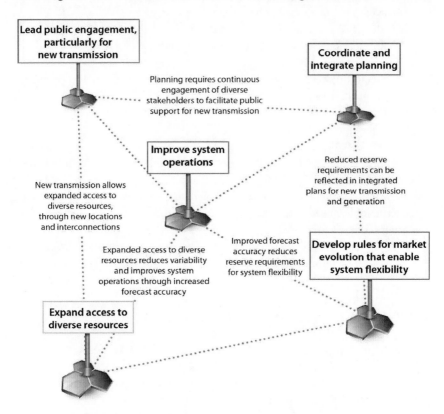

Figure 2. Example of interrelationships among areas of intervention.

Several studies, including the Eastern Wind Integration and Transmission Study (NREL 2011), the European Wind Integration Study (EWIS 2010), and the Greennet Study (summarized in Holttinen et al. 2009) have examined integration costs. The Eastern Wind Integration and Transmission Study found that among various scenarios, the interconnection-wide costs excluding transmission costs for integrating large amount of wind were less than $5 per megawatt-hour (MWh). The European Wind Integration Study examined both costs and benefits of incorporating high penetrations of wind, finding that the costs of managing the variability of wind ranged from €2.1to €2.6 ($2.7 to $3.4) per MWh, less than 5% of the calculated wind energy benefits.

The Greennet study estimated wind power balancing costs in Denmark at 28% market share amounted to €1-€2 ($1.3-$2.6)/MWh (Holttinen et al. 2009). The estimated values are similar to real world experience in West Denmark, where costs have been €1.4-€2.6 ($1.8-$3.4)/MWh at 24% wind penetration. These real world costs actually overestimate the balancing costs because the system operator collects more revenues than are actually required to pay the balancing resources used by the system, but nevertheless they are indicative of the magnitude of these costs. The Greennet study determined that additional balancing costs in Germany, at around 10% penetration, would be around €2.5 ($3.3)/MWh (Holttinen et al. 2009).

Conclusion

The cases reviewed for this analysis illustrate considerable diversity, not only of the electricity systems—and their supporting markets, institutions, and renewable resources—but in the actions each country has taken to effectively integrate high penetrations of variable RE. The cases reveal that there is no one-size-fits-all approach; each country has crafted its own combination of policies, market designs, and system operations to achieve the system reliability and flexibility needed to successfully integrate RE. Notwithstanding this diversity, the approaches coalesce around five strategic areas of intervention:

- Engage with the public, particularly in developing new transmission
- Optimize features of the power system over a broad geographic area through system wide comprehensive planning and the use of markets
- Adopt market designs that help support system flexibility
- Expand the diversity of resources—both in type and through expanded effective balancing areas
- Improve system operations, including integration of advanced forecasting to reduce the impact of RE variability, and grid codes that ensure system reliability.

The best practices associated with these five strategic areas benefit all power systems, not just those with high penetrations of variable RE. Yet these strategies are particularly instrumental in accommodating variable renewables where they minimize the impact of RE's variability and allow more options to cost-effectively strengthen the ability of a power system to respond to change. Advancements in energy efficiency and smart grids, when conjoined with higher RE integration, further strengthen the efficacy of any power system.

The study also emphatically underscores the value of countries sharing their experiences. Any country's ability to successfully integrate variable RE depends on a wide array of factors— technical requirements, resource options, planning processes, market rules, policies and regulations, institutional and human capacity, and what is happening in neighboring countries. The more diverse and robust the experience base from which a country can draw, the more likely that it will be able to implement an appropriate, optimized, and system-wide approach. This is as true for countries in the early stages of RE integration as it is for countries that have already had significant success. Going forward, successful RE integration will thus depend upon the ability to maintain a broad ecosystem perspective, to organize and make available the wealth of experiences, and to ensure that there is always a clear path from analysis to enactment.

References

Argus. (2011). "German Government Proposes Changes to CHP Subsidies." Accessed April 13, 2012: http://www.argusmedia.com/pages/NewsBody. aspx?id=778733&menu=yes.

Bird, L.; Milligan, M. (forthcoming). "WREF 2012: Lessons from Large-scale Renewable Energy Integration Studies." World Renewable Energy Forum.

BMU. (2011). "The Energy Concept and its Accelerated Implementation." (2011). Accessed April 13, 2012: http://www.bmu.de/english /transfor mation_of_the_energy_system/resolutionsandmeasures/doc /48054. php.

BMU. (2012). "Energiewende auf gutem Weg: Zwischenbilanz und Ausblick." [Energy Concept for an Environmentally Sound, Reliable and Affordable Energy Supply] Berlin: Bundesministerium für Umwelt, Naturschutz und Reaktorsicherheit (BMU) and Bundesministerium für Wirtschaft und Technologie (BMWi). http://www.bmu.de/files/pdfs/ allgemein/application/pdf/broschuere energiewende.pdf.

Burke, D.; O'Malley M. (2010). "Maximizing Firm Wind Connection to Security Constrained Transmission Networks." *IEEE Transactions on Power Systems* (25:2); pp. 749–759.

Cochran, J.; Bird, L.; Heeter, J.; Arent, D.J. (2012). "Integrating Variable Renewable Energy in Electric Power Markets: Best Practices from International Experience: Summary for Policymakers." NREL/TP-6A20-TP53730. Golden, CO: National Renewable Energy Laboratory."

Danish Electricity Infrastructure Committee (*Elinfrastrukturudvalget*). (2008). "Technical report on the Future Expansion and Undergrounding of the Electricity Transmission Grid: Summary." https://selvbetjening.preprod. energinet.dk/NR/rdonlyres/CC966C3A-FE78-41D8-9DC7-6B455210B502/0/TechnicalReportSummary.pdf. Accessed April 16, 2012.

Danish Energy Agency. (2011). Energy Statistics 2010. http://www.ens.dk/en-US/Info/FactsAndFigures/Energy_statistics_and_indicators/Annual%20St atistics/Documents/En ergy%20in%20Denmark%202010.pdf. Accessed April 16, 2012.

[DCENR] Department of Communications, Energy and Natural Resources and Department of Enterprise, Trade and Investment (DETI). (2008). "All Island Grid Study." http://www.dcenr.gov.ie/Energy/North-South+Co-operation+in+the+Energy +Sector/All+Island+Electricity+Grid+Study.htm.

Delille, G.; Francois, B.; Malarange, G. (2010). "Dynamic Frequency Control Support: a Virtual Inertia Provided by Distributed Energy Storage to Isolated Power Systems." *Proc. IEEE Innovative Smart Grids Technologies Europe Conference*, Gothenburg, Sweden, October.

[Dena] Deutsche Energie-Agentur GmbH. (2011). Grid Study II – Integration of Renewable Energy Sources in the German Power Supply System from 2015 – 2020 with an Outlook to 2025. Berlin, Germany: Dena. http://www.dena.de/fileadmin/userupload/Publikationen/Sonstiges /Dokumente/dena Grid Study II - final report.pdf.

Doan, L. (2010). "Texas Permits Routes for Contested CREZ Lines, Offers Glimpse of Future Decisions." *SNL Energy*. Subscription required. http://www.snl.com/interactivex/article.aspx?id=11008499&KLPT=6. Accessed March 28, 2012.

EirGrid; SONI. (2009). "TSO Facilitation of Renewables Studies." http://www.eirgrid.com/media/Facilitation%20of%20Renewables%20For um%20Session%201.pdf.

ERCOT (2008). ERCOT operations report on the EECP event of February 26, 2008. http://www.ercot.com/meetings/ros/keydocs/2008/0313/07. ERCOT OPERATIONS REPORT _EECP022608_public.doc Accessed February 2, 2012.

[EWIS] European Wind Integration Study. (2010). "Towards a Successful Integration of Large Scale Wind Power into European Electricity Grids. Final Report." Winter, W., ed. Brussels: ENTSO-E. Accessed April 17, 2012: http://www.wind-integration.eu/downloads/library/ EWIS Final Report.pdf.

Grant, W.; Edelson, D.; Dumas, J. Zack, J.; Ahlstrom, M.; Kehler, J.; Storck, P.; Lerner, J.; Parks, K.; Finley, C. (2009). "Change in the Air: Operational Challenges in Wind Power Production and Prediction," *IEE Power and Energy Magazine* (7:6); pp. 47–58.

Holttinen, H; Meibom, P.; Orths, A.; van Hulle, F.; Lange, B.; O'Malley, M.; Pierik, J.; Ummels, B.; Tande, J.O.; Estanqueiro, A.; Matos, M.; Gomez, E.; Söder, L.; Strbac, G.; Anser Shakoor; Ricardo, J.; Smith. J.C.; Milligan, M.; Ela, E. (2009). "Design and Operation of Power Systems with Large Amounts of Wind Power. Final report, IEA WIND Task 25, Phase One 2006–2008." Vuorimiehentie, Finland: VTT. http://www.vtt.fi/inf/pdf/tiedotteet/2009/T2493.pdf.

Holttinen, H.; Orths, A.; Eriksen, P.; Hidalgo, J.; Estanqueiro, A.; Groome, F.; Coughlan, Y.; Neumann, H.; Lange, B.; van Hulle, F.; Dudurych, I. (2011). "Balancing Act." *IEEE Power & Energy* (9:6); pp. 47–59.

Hur, K.; Boddeti, M.; Sarma, N.; Dumas, J.; Adams, J.; Chai, S. (2010). "High-wire Act." *IEEE Power & Energy Magazine* (8:1); pp.37–45.

[IEA] International Energy Agency. (2011). Harnessing Variable Renewable: A Guide to the Balancing Challenge. Paris, France: IEA.

[IPCC] Sims, R. Mercado, P.; Krewitt, W.; Bhuyan, G. ; Flynn, D. ; Holttinen, H. ; Jannuzzi, G.; Khennas, S.; Liu, Y.; O'Malley, M.; Nilsson, L.J. ; Ogden, J.; Ogimoto, K.; Outhred, H.; Ulleberg, Ø.; van Hulle, F. (2011). "Integration of Renewable Energy into Present and Future Energy Systems." Chapter 8. Edenhofer, O.; Pichs-Madruga, R.; Sokona, Y.; Seyboth, K.; Matschoss, P.; Kadner, S.; Zwickel, T.; Eickemeier, P.; Hansen, G.; Schlömer, S.; von Stechow, C. eds. *IPCC Special Report on Renewable Energy Sources and Climate Change Mitigation*, Cambridge, United Kingdom and New York, NY: Cambridge University Press; pp. 609-706.

Jewell, W.; Grossardt, T.; Bailey, K.; Gill, R.S. (2009). "A New Method for Public Involvement in Electric Transmission-line Routing." *IEEE Transactions on Power Delivery* (24:4); pp. 2240– 2247.

Jones, L.E. (2011). "Strategies and Decision Support Systems for Integrating Variable Energy Resources in Control Centers for Reliable Operations:

Global Best Practices, Examples of Excellence and Lessons Learned."
Washington, DC: Alstom Grid Inc.

King, J.; Kirby, B.; Milligan, M.; Beuning, S. (2011). "Flexibility Reserve
Reductions from an Energy Imbalance Market with High Levels of Wind
Energy in the Western Interconnection." NREL/TP-5500-52330. Golden,
CO: National Renewable Energy Laboratory. http://www.nrel.gov/docs
/fy12osti/52330.pdf.

Kirby, B.J. (2007). "Load Response Fundamentally Matches Power System
Reliability Requirements." *Power Engineering Society General Meeting,
IEEE*, Tampa, FL, USA, 2428 June 2007, pp.1–6.

Kiviluaoma, J.; Meibom, P. (2010). "Influence of Wind Power, Plug-in
Electric Vehicles, and Heat Storages on Power System Investments."
Energy (35); pp. 1244–1255.

Lasher, Warren. (December 2, 2011). Phone interview. Electric Reliability
Council of Texas. Austin, TX.

Lauby, M.; Ahlstrom, M.; Brooks, D.; Beuning, S.; Caspary, J.; Grant, W.;
Kirby, B.; Milligan, M.; O'Malley, M.; Patel, M.; Piwko, R.; Pourbeik, P.;
Shirmohammadi, D.; Smith, J. (2011). "Balancing Act." *IEEE Power &
Energy* (9:6); pp. 75–85.

Milligan, M.; Ela, E.; Hodge, B.M.; Kirby, B.; Lew, D.; Clark, C.; DeCesaro,
J.; Lynn, K. (2011). "Cost-causation and Integration Cost Analysis for
Variable Generation." *Electricity Journal* (24:9).

Milligan, M.; Kirby, B. (2010a). *Market Characteristics for Efficient
Integration of Variable Generation in the Western Interconnection*. NREL
Report No. TP-550-48192.

Milligan, M.; Kirby, B. (2010b). *Utilizing Load Response for Wind and Solar
Integration and Power System Reliability*. NREL/CP-550-48247. Golden,
CO: National Renewable Energy Laboratory. http://www.nrel.gov/docs/fy
10osti/48247.pdf.

Netzentwicklungsplan [Network development plan] Gas 2012. Accessed
February 20, 2012: http://www.netzentwicklungsplan-as.de/files/entwurf
netzentwicklungsplan gas 2012.pdf

[NREL] National Renewable Energy Laboratory. (2011). *Eastern Wind
Integration and Transmission Study*. NREL/ SR-5500-47078. Work
performed by EnerNex Corporation, Knoxville, TN. Golden, CO: National
Renewable Energy Laboratory. http://www.nrel.gov/docs/fy11osti/47078.
pdf.

NREL. (2010). *Western Wind and Solar Integration Study*. NREL/SR-550-
47434. Work performed by GE Energy, New York, NY. Golden, CO:

National Renewable Energy Laboratory. http://www.nrel.gov/docs/fy
10osti/47434.pdf.

[NERC] North American Electric Reliability Corporation. (2011). "IVGTF
Task 2.4 Report: Operating Practices, Procedures, and Tools." Princeton,
NJ: NERC, March. http://www.nerc .com/docs/pc/ivgtf/IVGTF2-
4CleanBK(11.22).pdf.

NERC. (2010). "Flexibility Requirements and Potential Metrics for Variable
Generation: Implications for System Planning Studies." Princeton, NJ:
NERC, August. http://www.nerc.com/ docs/pc/ivgtf/IVGTF_Task_1_4_
Final.pdf.

NERC. (2009). "Accommodating High Levels of Variable Generation."
Princeton, NJ: NERC, April. http://www.nerc.com/files/IVGTF_Report_
041609.pdf.

REALISEGRID. (2011). *Improving Consensus on New Transmission
Infrastructures by a thorough Presentation of the Benefits Given by
Priority Projects.* Work performed by Technische Universiteit Delft.
http://realisegrid.rse-web.it/content/files/File/Publications%20and%20
results/Deliverable REALISEGRID 3.7.2.pdf.

Reid, Rob. (February 23, 2012). Phone interview. POWER Engineers. Austin,
TX.

REN21. (2011). Renewables 2011 Global Status Report. Paris: REN21
Secretariat.   http://www.ren21.net/Portals/97/documents/GSR/REN21
GSR2011.pdf

TenneT (2012). "TenneT Structural Solution Facilitates German Energy
Transition." Accessed April 13, 2012: http://www.tennet.org/english
/tennet/news/tennet-structural-solution-facilitatesgerman-energy-
transition.aspx.

VTT 2007: Optimising the Market Integration of Electricity from Renewables;
Holttinen, H., Korenoff, G., Lemström, B., Helsinki, Finland, December.

Wattles, P. (December 8, 2011). Personal communication and email. Electric
Reliability Council of Texas. Austin, TX.

[WECC] Western Electricity Coordinating Council. (2011). "WECC Efficient
Dispatch Toolkit Cost-benefit Analysis." Salt Lake City, UT: WECC.
http://www.wecc.biz/committees/EDT/EDT%20Results/EDT%20Cost%
20Benefit%20Analysis%20Report%20-%20REVISED.pdf.

APPENDIX A. INTRODUCTION TO THE CASE STUDIES

The following appendices present case study examples from six countries (Australia, Denmark, Germany, Ireland, Spain, and the United States) that have seen high penetrations of renewable energy. Appendix A provides an overview of the electricity mix in each country. Each country has a unique energy mix; of the countries examined, Denmark, Spain and Ireland saw the greatest percentages of wind generation in 2010, while Germany and Spain had the greatest percentages of solar, tide and wave generation (Figure A-1).[17]

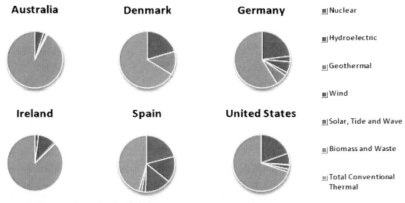

Source data are from EIA (2012).

Figure A-1. Percentage of electricity generation by type, 2010.

Data from 2010 provide only a snapshot in time. Figures A2 through A7 present growth in non-hydropower renewable generation from 2001 to 2010. Each case study country has seen different levels of non-hydro generation. In Denmark, Germany, and Spain, solar, tide and wave energy saw the largest growth rates from 2001 to 2010, though other sectors (wind or biomass and waste) contributed most on a generation basis in 2010. In Australia, Ireland, and the United States, the wind sector saw the greatest growth in percentage terms, and wind also contributed the most generation in 2010.

From 2001 to 2010, wind generation in Australia increased from 0.2 billion kWh to more than 3.6 billion kWh, representing a compound annual growth rate of 33% (Figure A-2).

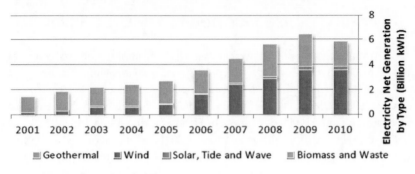

Source data are from EIA (2012).

Figure A-2. Non-hydropower generation (billion kWh) in Australia, 2001-2010.

In Denmark, wind has continued to play a strong role, and biomass and waste resources increased from 2.0 billion kWh in 2001 to 4.8 billion kWh in 2011 (Figure A-3).

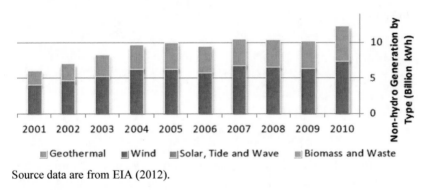

Source data are from EIA (2012).

Figure A-3. Non-hydropower generation (billion kWh) in Denmark, 2001-2010.

In Germany, solar has seen the largest growth from 2001 to 2010 (Figure A-4), though there has been strong growth in wind as well. The share of PV increased from 0.1% in 2003 to 1.9% in 2010, representing 11,683 GWh (BMU 2011, Eurostat 2011).

Ireland's renewable energy development is almost exclusively wind, which has seen compound annual growth rate of 24% from 2001 to 2010 (Figure A-5).

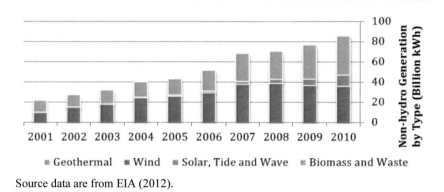

■ Geothermal   ■ Wind   ■ Solar, Tide and Wave   ■ Biomass and Waste

Source data are from EIA (2012).

Figure A-4. Non-hydropower generation (billion kWh) in Germany, 2001-2010.

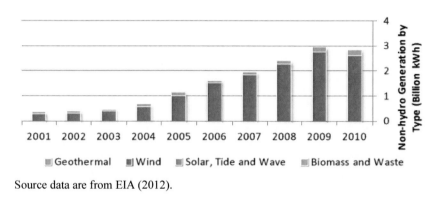

■ Geothermal   ■ Wind   ■ Solar, Tide and Wave   ■ Biomass and Waste

Source data are from EIA (2012).

Figure A-5. Non-hydropower generation (billion kWh) in Ireland, 2001-2010.

Spain provides a second example of where, in addition to wind, the solar sector has seen large growth (Figure A-6). Solar increased from 0.02 billion kWh in 2001 to 6.3 billion kWh in 2010, representing compound annual growth rate of 75%.

In the United States, the greatest growth has come from wind generation, which increased from 6.7 billion kWh in 2001 to more than 94 billion kWh in 2010 (Figure A-7).

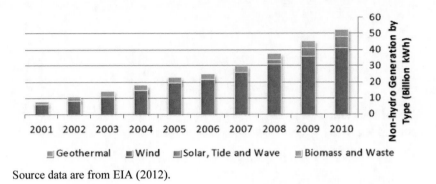

Source data are from EIA (2012).

Figure A-6. Non-hydropower generation (billion kWh) in Spain, 2001-2010.

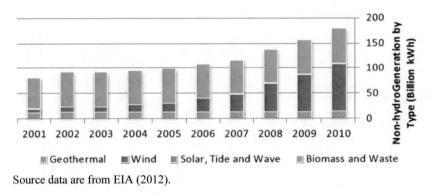

Source data are from EIA (2012).

Figure A-7. Non-hydropower generation (billion kWh) in the United States, 2001-2010.

References

BMU (Federal Ministry for the Environment, Nature Conservation and Nuclear Safety) (2011).

Renewable Energy Sources in Figures, National and International Development. Berlin. July. Eurostat (2011). Online Eurostat database. Accessed 30th January 2011.

EIA (Energy Information Administration) (2012). International Energy Statistics. http://205.254.135.7/cfapps/ipdbproject/IEDIndex3.cfm? tid= 2&pid=2&aid=12 Accessed April 2, 2012.

APPENDIX B. CASE STUDY: AUSTRALIA (HIGH WIND ENERGY PENETRATION IN SOUTH)

Author: Craig Oakeshott, Principal Consultant,
Sinclair Knight Merz (SKM)
Editor: David Swift, Executive General Manager,
Australian Energy Market Operator

Background

The National Electricity Market (NEM) has been operating continuously in Australia since December 12, 1998. The NEM covers 92% of Australian electricity demand, and provides a common market structure, under the National Electricity Law (NEL), and single market operator, the Australian Energy Market Operator (AEMO), for the interconnected states: Queensland, New South Wales, Victoria, Tasmania, and South Australia. Generators, Transmission Network Service Providers (TNSP), Distribution Network Service Providers (DNSP) and retailers all have defined, non-overlapping, roles in the NEM and may, or may not, be privately owned. The TNSPs and DNSPs receive a regulated rate of return for the cost of operating the network and on investments that pass a market benefits test. Network fees are charged primarily to customers as part of their retail energy price. The NEM design is technology neutral. All generation is dispatched based on its market offers. The NEM offers open access to generators and, as such, there is no central planning directing generator investment in the NEM. Rather the size, configuration, fuel source, timing, and operational role of all generator investments are made by the proponent based on its own economic assessment. Market dispatch and network congestion provide pricing signals. Enhanced locational incentives within a region are provided by modifying regional prices with network losses and by imposing the cost of connection to the network on the proponent. Renewable energy targets and subsidies all act to change the economics of the decision to invest, but these occur independently of the energy market. See AEMO (2010) for a broad introduction to the NEM.

The NEM has been characterised by its clear governance structures and a strong adherence to sound market principles. South Australia has been part of the NEM since its inception and has a low population density. A number of long term wind studies in the mid 1980s showed that South Australia had a very attractive wind resource and provided a guide for identifying good locations. The introduction of a growing renewable generation obligation on retailers in the NEM from 2001 triggered the start of grid connected wind generation in South Australia. This obligation on retailers, to purchase a regulated percentage of renewable energy, provides an additional revenue stream to qualifying generators.

Table B-1 lists the population and key energy statistics for South Australia. Figure B-1 shows percentages of installed capacity in South Australia by generation type.

Table B-1. Statistical Summary of Population and Energy for South Australia

South Australian population (30 June 2011)[a]	1,657,000
Customer sales (Estimate for 30 June 2011)	13,045 GWh
Peak demand (actual uncorrected 30 June 2011)	3385.42 MW
Average demand (actual uncorrected 30 June 2011)	1543.04 MW
Load factor	45.57%
Wind installed and operating at 30 June 2011	1150.35 MW
Wind energy production per capita	1807 kWh/person
Installed wind capacity per capita	694 W/person

[a] Australian Bureau of Statistics (2011).

Coordinate and Integrate Planning

The current electricity market planning arrangements were initiated following a review by the Council of Australian Governments. Their decision significantly enhances the original arrangement, recognising that the original jurisdictionally based approach was unlikely to deliver a truly national examination of grid expansion options, or result in the most economic national transmission network development. However, national planning remains an information resource only. Network planning and development is focussed primarily on reliably meeting customer demand as opposed to providing transfer capacity to support generators or meet renewable energy targets. While the majority of network investment has focussed on meeting the reliability standards, the planning approach does assess the expected future level of constraint in the network and does identify augmentations that deliver net market benefits.

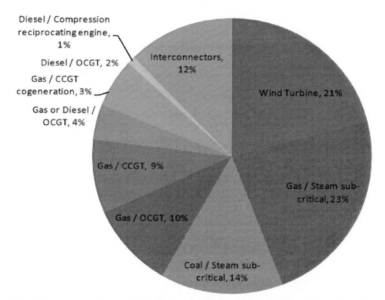

Calculated from statistics published in the AEMO 2011 Electricity Statement of Opportunities.

Figure D-1. Percentage of installed capacity in South Australia, by technology.

The NEM provides a competitive wholesale electricity generation market and open access to the transmission network. Subsidies and targets for renewable energy affect the economics of investment decisions made by the generators but do so outside NEM operation and settlement. Investors may develop new generation projects at any location, time, and with any technology in response to their interpretation of market signals. The planning process may inform the investor of issues or opportunities, but there is no mechanism to force any particular outcome and any decision to invest in new generation is entirely the responsibility of the investing party. Most of the generators in the NEM, and particularly the renewable energy generators, are privately owned. Investments therefore will be driven by commercial imperatives, and they should be viable although they may not be the most economically efficient outcome from a national perspective.

AEMO publishes a range of reports that seek to inform the market about potential opportunities for generation investment and network development, now, and into the future.

AEMO's annual *Electricity Statement of Opportunities* (2011) lists current operating plant, new projects that are at different stages of development, and a forecast of the year in the future when the anticipated load growth will exceed the level of available capacity needed to maintain the market reliability standard.

AEMO's National Transmission Network Development Plan (NTNDP) specifically delivers a forecast of the network and market development over the next 20 years. The plan seeks to deliver the lowest-cost solution by optimising the expansion of both the network and generation and the mix of existing and future technologies employed. Integrating renewable generation into this planning process has been, thus far, relatively simple. The NTNDP analysis uses market simulation tools that incorporate reasonable cost estimates for all technologies. These costs are independently researched and verified in consultation with industry. The carbon price and the value of renewable energy certificates are applied to the resource costs and they therefore affect the relative costs between technologies over time.

While the generation development path from the NTNDP is only indicative, it highlights regions of the NEM where development is expected to be the most economically viable. The NTNDP informs participants about the potential generation mix in each state in the NEM, a view of network constraints and the potential lowest-cost network development path. However, project developers are driven by many factors that cannot be modelled, and they may not always choose the most economically efficient locations or configurations. The 2011 NTNDP explored the operational and constraint challenges that may arise if the significant levels of renewable energy generation in the NEM that were forecast in the 2010 NTNDP occur. [18] The NTNDP also provides the Australian Energy Regulator with an independent reference on future market development to assist in its examination of regulated network development proposals from the TNSPs.

While development of intermittent[19] renewable generation, such as wind, could be aided by a central planning body developing large-scale transmission to areas with suitable resources, in an environment where most of the players are commercial organizations, such a decision would require "picking winners". Project proponents pay for the connection assets between their plants and the shared network, but all customers pay for the cost of the shared transmission network and its development. Justifying which states, what locations and which resources should be developed would be problematic, as it would always result in winners and losers in the generation sector. At this stage, the total capacity of the renewable energy projects being discussed exceeds the existing targets, and there has been no shortage of projects that could be effectively developed close to the existing network. There is also at least one, privately proposed, large-scale transmission project being considered that would facilitate significant remote renewable generation and load development. The market benefits of these projects are yet to be assessed. Proponents can elect to pay for the removal of a constraint or partially fund an augmentation to the shared network. However, at this stage, the network capacity made available by such an investment is still subject to open access, and could be used by any participant. Generation investors cannot obtain rights over transmission capacity.

Development of capacity in the shared network relies on the TNSP, independently or in response to enquiries from a proponent, identifying a potential need and follows a formal justification process. This formal process is a market benefits test, known as the Regulatory Investment Test – transmission (RIT-t)[20]. RIT-ts are performed for each project to identify the solution that provides the maximum market benefits by examining appropriate transmission, generation, and demand-side projects that could relieve the network constraint. Costs and benefits considered in the test include reliability and potential competition benefits as well as total fuel, capital, operating and maintenance costs for all plant and equipment in the network.

Amendments to the NEM were proposed to enable a group of proponents to request a network solution to deliver power from a cluster of generation projects distant from the shared network and to realise the potential economies of scale through developing a common facility. Under the proposal, TNSPs would initially fund the required infrastructure for the proponents by an increase in transmission charges on customers. As proponents actually connect to the new assets, each would repay its share of the total cost of funding, offsetting future charges. This proposal was not ultimately implemented and the task of developing of a suitable alternative has been included in the current Transmission Framework Review.

A Transmission Framework Review is currently underway to address potential deficiencies in the longer-term development of the national transmission system. This review is designed to address perceived barriers to investment in transmission, national cost/benefit allocation between states, improved mechanisms to resolve network congestion and cost allocation for remote transmission development. These initiatives could improve the clarity of economic signals for development without compromising the market objective or promoting specific technologies and targets.

Community Support and Land Use Planning by State Governments and Local Councils

South Australia has large areas of land suitable for renewable energy deployment and relatively sparse population outside of the capital city. While surveys show a majority of the community supports renewable energy development, some incorporated groups now oppose development in more populated areas.

The land use planning approval processes, run by local councils and state governments, includes community engagement and feedback on the acceptability of specific developments. Proponents can obtain planning approval from local government or obtain state sanctioned special project status and a streamlined approval process through state government. Large wind farm developers can choose either approach, and there have been successes and failures for both alternatives. Ultimately, the approvals process can become highly politicised, and it may be governed by a range of issues in addition to local concerns. The Australian Government, while the architect of the renewable energy target, has no control over development approvals.

Public information from planning bodies, such as the South Australian Electricity Supply Industry Planning Council[21] (ESIPC), informed the wind developers of wider state and NEM issues through dedicated public forums and the annual planning publications. These mechanisms provided the public, proponents, governments, and regulators independent analysis of market opportunities, operational requirements, and responsibilities with respect to development costs. For example, the ESIPC presented information on the balance between the project connection costs and the available wind resource. This highlighted the advantage of building wind farms in areas with existing transmission capacity and a good wind resource, rather than in areas with exceptional wind resources but no grid.

Gaining acceptance of wind development by the community remains a challenge for developers in South Australia and nationally. Active community engagement, through public fora and sponsored events in communities affected by the development of new wind farms are used to garner local support. Landholders that have turbines located on their properties are typically provided annual compensation per turbine. While long-term local employment opportunities are potentially limited to a few individuals, the additional money circulating in the community has many long-term benefits.

On a larger scale, many of the major developers have run media campaigns to publicize the environmental benefits of their renewable energy activities and to promote their green energy products to consumers. Concerns about visual amenities have led to the development of new state government guidelines for the siting of wind farms with respect to local communities and housing.

Many new renewable energy generation technologies are being investigated in areas where the renewable energy resource is rich but there is no transmission network. While the South Australian and the Australian Government may provide some seed capital and development incentives, no government in Australia to date is offering to build transmission infrastructure with public funds. This approach of not "picking winners" is designed to maintain a level playing field for development and drive projects with the lowest overall costs first.

A number of published studies have investigated the cost of transmission to deliver renewable energy from remote resources and potential additional benefits from augmented interconnection between NEM regions. At the core of these assessments is the comparison between the cost of generating and delivering energy from a higher quality remote resource to the cost of generating and delivering energy from a poorer resource close to the load centre.

Lessons Learned

- A well-designed framework for government land use approvals provides a streamlined and efficient process to balance community issues with project development that facilitates development and manages local support.
- Maintaining community support, particularly in areas where the wind resources are good, is an ongoing issue and needs to be actively managed through direct community involvement.
- The regional pricing and congestion signals calculated in the NEM and the cost of connection on proponents have encouraged development in areas well serviced by transmission capacity and not in areas where major transmission augmentation and connection costs would be required.
- A market benefits test assessment of transmission augmentations ensures that consumers pay only for developments.
- Analysis shows that there may be efficiency advantages if suitable schemes can be developed to realise the economies of scale possible in connecting new generation clusters remote from the grid.

Enable Markets to Facilitate Variable Generation

While some minor modifications to the NEM objective have occurred, it has remained effectively constant since the inception of the market and throughout the period of renewable energy development. The market objective is:

> "to promote efficient investment in, and efficient operation and use of, electricity services for the long term interests of consumers of electricity with respect to – price, quality, safety, reliability, and security of supply of electricity; and the reliability, safety and security of the national electricity system." (AEMO n.d.)

Market operation is based on the premise of simplicity and transparency.

Key principals of the market operation that have remained constant since the inception of the market are:

- Gross pool with mandatory participation of all larger generators (> 30MW)
- Energy-only market based on security constrained optimised dispatch
- Energy dispatch and pricing co-optimised with market ancillary services
- Self commitment market with 5-minute dispatch on the basis of the latest offers
- Offers can be changed any time prior to dispatch
- 5-minute regional clearing price, 30-minute settlement period
- Dispatch optimisation includes impacts of network constraints and capability
- Network constraints affect price
- Single regional market price referenced to each generator's connection point by the inclusion of losses
- Market price is collared (-$1030/MWh to $12,870/MWh) where negative prices provide strong incentives against oversupply
- Single market operator directly controls all significant generation on the interconnected network, allowing the optimal use of the power system whilst maintaining system security
- The market has no cross-temporal optimisation but a great deal of forward information with respect to forecast prices and demand is provided to participants to assist their planning in every timeframe

No wind farms existed in Australia on the shared network at the start of the market in 1998 and no consideration had been given to this form of generation making a significant contribution to the future energy needs of the customers.

The rules of the market have been amended to incorporate intermittent generation but in a manner that is consistent with the market objective and the existing market model. The necessary changes were determined by evidence-based analysis, in consultation with all participants in the industry and with the support of, but independent from governments. Investments made in the past were not been required to meet new requirements but rather were been grandfathered.

Integrating Intermittent Generation

Integrating wind generation into the operation of the market focused on technology neutrality and ensuring power system security, especially security against the risk of cascading failures.

The original market design considered only two dispatch categories: "scheduled" and "nonscheduled" generation. Wind was initially registered in the "non-scheduled" category, and consequently its output was not subject to central control or optimization, leading potentially to stability and security issues. A third category, "semi-scheduled" generation, was established for intermittent generation that maintained the obligations for control and the information provisions from the "scheduled" category but was tailored to better match the character of intermittent generation and to facilitate the integration of wind into market systems. This included providing a verified energy conversion model to convert forecast wind speed into power.

Generators in the semi-scheduled category submit offers for their output (dollars for megawatts dispatched) and are included in constraints and market dispatch. Forecasts of the output of each wind farm are incorporated into the market systems, and if a wind farm is part of a binding network constraint, depending upon their bid price compared to other generators in that constraint, its output can be capped. Forecasts of the total regional wind contribution, customer demand projections, and potential network constraints are provided to all market participants. Other forms of generation are therefore aware of the opportunities and can make their own decisions as to when to commit plant and the price and quantity of generation to offer.

Wind generation, as a semi-scheduled generator, also pays a share of ancillary service costs calculated on a "causer pays" basis. This calculation summates the contribution each generator makes to the need for system frequency control.

Lessons Learned

- A stable market design that can evolve to help integrate variable renewable generation has reduced sovereign risk and provided some security for existing and proposed developments. Providing a simple, clear and transparent market-operating environment from which market signals are evident is imperative and significantly aids integration.

Motivate System Flexibility

The NEM provides incentives for greater flexibility in the behaviour of generators. As the proportion of wind generation in the South Australian region has grown, the frequency of low and negative pool prices has increased. Low prices occur when there is significant surplus of generation, such as when wind output is high and demand is at average or low levels. This has provided:

- Incentives for wind generation to reduce output in periods of negative prices, assisting through market signals to resolve a potential situation of excess generation
- Incentives for other plant to increase its flexibility with changes to control systems and operating arrangements to allow generators to optimise their behaviours to match the opportunities in the market
  - In South Australia, older coal fired generators with steam bypass capabilities are now using these facilities to reduce their output to levels below their normal minimum stable operational point to avoid negative prices during periods of excess supply.

The NEM is an energy-only market with close to real-time pricing. While prices in South Australia can be low, even negative, when supply is high relative to demand, prices can also rise quickly when wind output falls and other generation needs to ramp up its output or start generating. The NEM has therefore also provided an incentive for new flexible, fast-start generators to enter the market allowing future investments to manage their risk and maximise their earnings potential.

Promoting Diversification of Location and Type of Renewables
The Australian approach has been to allow market signals to drive new investment to the extent possible. All of the developers of new generation projects in Australia are privately owned, and their investment decisions are likely to be based on delivering commercial rates of return and should account for the location, technology, timing, and anticipated market role of the proposed generator.

The NEM is a regional market and regional wholesale market electricity prices are set, as part of the dispatch process, in every dispatch interval. The price at every connection point is derived from the regional reference price modified by a locational-specific multiplier that reflects the incremental network losses to that point. The multiplier, the cost of connection to the network, and the likelihood of constraints in the network all act as signals for new generation investments. These are all part of the normal operation of the market and are not specific to a generation technology. Market participants are well aware of these factors and are careful to understand the potential exposure that their project may have during the feasibility stage of development.

Regional Pricing
Each state or region in the NEM has a nominal "load centre" known as the Regional Reference Node. The market dispatch process establishes a regional price at each regional reference node in every dispatch interval. The performance of each nominal interconnector between the regional reference nodes is represented by a mathematical equation. These mathematical equations are used in the dispatch process to rank the relative price of an additional megawatt of capacity over the interconnector as compared to one generated locally. Over time, generation investment will be biased toward those regions with the highest price and hence greatest need for generation.

- Based on discussion with the wind proponents, the occurrence of negative prices in South Australia clearly factor into their decisions for ongoing investment. The frequency of low prices is also affecting the forward contract prices for both wind farms and fossil-fuelled generation.
- Wind proponents are not advancing their projects until power purchase contracts have been finalized.
- While the number of wind farms under consideration is still considerable, a number of conventional open cycle gas turbine generation projects are also progressing and the transmission network

service provider is actively investigating additional interconnector capacity.

The pattern of half-hourly prices over a year provides a very sophisticated price signal. The generation weighted average price for any particular generation pattern reflects the earnings for that generator. South Australia experiences a relatively uniform wind regime across the state. As wind generation in the state has grown, the market earnings of those wind generators has fallen in relative terms. During periods of high wind, the wholesale market price is significantly depressed because of the relative abundance of supply. Consequently, wind farm earnings during the periods when their output is the greatest is low. During calm periods, the relative scarcity of supply is higher and market prices are higher. It can be clearly shown that wind generators in South Australia earn a lower volume-weighted average price than other generation in the region and as the penetration of wind has increased the gap between the volume and time weighted prices received by the wind farms compared to fossil fired generation has grown.

Table D-2 highlights the growing difference between these earnings prices.

Table B-2. Time and Volume-Weighted Prices for Renewable and Fossil-Fuelled Generation in South Australia ($A/MWh)

Financial Year	Renewables		Fossil Fuels		South Australian Market		Full year Time-Weighted ($A/MWh)
	Full Year ($A/MWh)	Summer ($A/MWh)	Full Year ($A/MWh)	Summer ($A/MWh)	Full Year ($A/MWh)	Summer ($A/MWh)	
2003/04	33.09	40.56	39.96	50.43	43.85	61.09	32.57
2004/05	38.47	56.72	44.56	67.5	44.61	67.92	33.33
2005/06	32.57	39.59	43.91	67.5	43.26	65.78	34.34
2006/07	49.69	51.55	58.71	67.21	58.35	66.43	62.95
2007/08	63.31	63.94	102.01	149.92	98.46	142.32	55.95
2008/09	46.39	91.8	70.5	165.34	67.16	155.12	38.76
2009/10	47.39	77.43	86.69	140.98	80.17	131.01	28.44
2010/11[a]	22.82	29.75	50.78	91.74	45.17	78.60	27.63

[a] Pro rata assessment for July 1, 2010 to 31 March 31, 2011.

The price signal can be seen to encourage diversification in location, wind resource, and technology of generators. South Australian wind is not as well correlated to peak demand periods. Developers are now considering wind generation in other regions of the NEM where the correlation of output to periods of higher demand is greater, increasing the potential market value of their output. There are also potential benefits to other forms of intermittent renewable generation that can generate when the wind output is lower. Solar power could offer such diversity and has a higher correlation to peak demand, which could be expected to improve the generation-weighted price it obtains. At this stage, despite potentially higher returns and the renewable energy credits, the only large-scale solar projects being actively pursued are in other states, and they rely on significant additional Australian government subsidies.

Marginal Loss Factors within a Region

Within a region, the marginal loss factor represents the losses between a connection point and the regional reference node. This is effectively a proxy for the relative balance between supply and demand at a location within the region with respect to the regional reference node. This signal is in the form of a multiplier to the regional reference price used in the dispatch process and in market settlements. Marginal loss factors for all generators and customers are calculated annually one year ahead based on known projects and forecast demand at each location.

Where local supply exceeds the proximal demand, the price received by the generators at that connection point is discounted to reflect the transmission losses incurred transferring the excess power to distant customers. The marginal loss factor discounts the market price received by the generators in that location providing a disincentive for further development at that location. Conversely, where local demand exceeds the proximal supply a marginal loss factor would provide an incentive for local supply investment. In a state such as South Australia, where there is such a large amount of suitable land with adequate wind resources, such a signal to relocate is not seen as limiting wind penetration. Rather, it provides a signal to explore appropriately sized opportunities close to load. Most developers incorporate this aspect into their investment decision.

Network

Developers negotiate with the transmission network service provider to determine the cost and capacity of their proposed connection. All generators are responsible for their "shallow" connection costs to the grid, providing an additional locational signal.

The negotiation of a satisfactory connection however, does not warrant any particular level of access to the shared grid. Network constraints, in the "shared network", between the proponent's project and potential demand centers may arise with new generation, especially in areas with good resources. In the NEM, congestion will be considered for removal but will not be eliminated unless an upgrade represents economically efficient solution to additional supply. This then drives developers to seek projects sited close to strong points in the network to minimise connection costs and congestion on the shared network.

Developing and Integrating Advanced Forecasting Techniques into Grid Operations

The energy-only market design in Australia relies on the market operator using a forecast of future customer demand, wind production, and advanced generator bids in a simulated market dispatch to provide a forward price projection up to seven days ahead to drive efficient outcomes. The NEM has no inter-temporal optimisation and does not pay start-up or shut down costs. Generators use these price forecasts to make their own decisions on when to commit their plant. Wind generation is part of this prediction.

It was recognised early in the development of wind power in South Australia that without a state-of-the-art forecasting system to inform the NEM of the potential contribution of wind power to system operation and security would be adversely affected. It was seen as unnecessarily onerous, inefficient, and potentially inaccurate for each wind farm to attempt to forecast its own output over market operational and planning time frames. A number of proponents were significantly concerned about their ability to fund and continuously operate individual forecasting systems that would be suitable for market operation. The ESIPC and the market operator initiated working groups to examine the potential for the development of a national system and engaged with the Australian Government as a potential source of funds. Together, these bodies established a model[22] to develop and deliver a suitable wind forecasting system, for all participants, that would be fully integrated into the market dispatch and pricing process. Wind generators are obliged to provide information to the market operator as inputs to the national wind forecasting system. As a centrally coordinated system it incorporates inputs from a wide range of public and private resources, from individual wind farm or turbine outputs and on-ground automatic weather stations to satellite imagery.

The system[23] delivers forecasts of wind production for all wind generators. These are published in aggregate for all participants with the demand forecast and individual production profiles are provided to the appropriate market participants in all market time frames.

The design of the forecasting system is such that it is also a platform for researchers to improve forecasting processes and will be able to accommodate other intermittent renewable generation, such a solar.

The greatest risk to security and reliability of supply of wind generation is the potential for rapid changes in output caused by sudden changes in wind speed or by wind speeds above cut-off speeds. The forecasting process uses the average of the predicted wind speed outcomes from a series of different atmospheric models and performs well under stable atmospheric conditions. However, where the individual forecasts diverge, particularly during periods where atmospheric conditions are likely to change rapidly, predicting the timing and magnitude of these rapid change events is less reliable. In recognition of this situation, research[24] into additional forecasting functionality was commissioned through the University of New South Wales and has delivered some potentially significant improvements in this area.

Lessons Learned
- A centrally funded and coordinated wind forecasting system has provided an integrated, reliable, scale-efficient, and world-best practice facility on which market dispatch systems and participants can depend.
- Integration of the wind forecasting system into market systems has aided decision making by all market participants and the market operator.
- A central, commonly supported forecasting service ensures consistency and extends the potential for providing advanced warning of rapid change events by integrating inputs from wind farms, automatic weather stations and other meteorological sources.
- The central facility can be expanded to accommodate other forms of renewable energy, providing an automatic reflection of new technologies into the market systems and dispatch processes and supporting industry research, supporting its longevity and constant improvement.
- Putting obligations on all wind participants has ensured the consistency of information and has reduced participant overheads

Supporting Model Grid Codes
Just as the original market dispatch arrangements were not well matched for high concentrations of intermittent renewables such as wind farms, the technical standards were also poorly suited to intermittent renewable energy

generators. The technical standards at the start of the NEM were designed around the capabilities of conventional gas, coal, and hydro-powered generation. The latest technical standard revisions have focussed on removing this technological bias and establishing more universal performance requirements.

Rapid development of wind farms in South Australia overtook national processes to adjust national technical standards. The South Australian Essential Services Commissioner and the state Energy Minister requested the ESIPC review the impact of wind generation on the price, reliability and power quality of wind generation. The ESIPC (ESIPC 2005) considered that a cap on installed wind capacity would be difficult to administer and inconsistent with the market philosophy, and it instead recommended a range of measures to integrate wind generation into the market, including higher technical standards for future wind farm connections.

All generators in South Australia must be licensed by the Essential Services Commissioner before they are permitted to operate. The technical standards recommended by the ESIPC were imposed as special license conditions on new wind farms. These standards were, to the extent possible, consistent with the national generator standards but biased toward the higher end of requirements. Specifically, in South Australia, the standards required all wind farms to:

- Register and operate as scheduled generators until the national "semi-scheduled" classification came into existence
- Participate in the national forecasting initiative that was just commencing at the time
- At the connection point to the shared network the wind farm must:
  - Ride through a 170 millisecond two phase to ground fault,
  - Provide voltage control,
  - Be able to operate at ± 0.93 power factor (50% of this to be dynamic fast acting; the other 50% to be slow acting and, with the consent of the TNSP, it could be differed to a time in the future when it may be needed, and
  - Provide full remote control capability for the local TNSP/market operator.

Significantly, the requirements focused on the performance of the wind farm at the connection point rather than on the turbines or individual components within the farm. It is particularly important that the reactive requirement be tested at the connection point. With equipment such as a STATCOM, a wind farm composed of less expensive, lower-performance turbines could comply with the requirements. The standards have delivered a stable and secure power system in South Australia with no recorded incidents despite occasions when more than 75% of the state's power was supplied by wind generation.

A review of the national Technical Standards provided greater flexibility for negotiation and incorporated some of the requirements identified by the ESIPC. Subsequently, the South Australian licence conditions were revised to align with specific clauses in the NEM technical standards but have not been significantly relaxed.

Opinion on these license conditions from the wind industry is polarised. Some proponents have praised their clarity and simplicity and have indicated that this mechanism has supported the ongoing development of the industry in the state by ensuring that each new wind farm will not be adversely affected by neighbouring farms. Others have indicated that the additional cost of compliance has adversely affected the financial viability of their projects. The ESIPC estimated that the additional cost of the license conditions added between two and three percent to the total capital cost of the wind farm projects.

The ESIPC's approach of requiring each wind farm to install reactive plant at its connection point has been criticised for delivering an oversupply of reactive capability in areas where a number of closely spaced farms have been built. However, the approach was considered justified in the absence of a more comprehensive regime, or market, for the provision of reactive power.

Lessons Learned
- Providing a level playing field that maximises the potential development by minimizing impacts of one generator on another is very important.
- Focussing on the performance at the connection to the network is more effective in managing that impact than predefining the performance of individual components.

Political Environment, Initiatives, and Incentives
The development activity based on Australia's renewable energy targets has been high and has benefited from bipartisan political support. A high level of policy certainty is clearly required to underpin market-based investment.

South Australia has a stable political environment with well-defined processes for gaining planning and development approval for all projects, not just electricity or energy. To date, these processes have applied to the development of wind generation in South Australia.

The national Mandatory Renewable Energy Target scheme imposed a requirement for retailers to purchase a given percentage of their energy from renewable generation sources. The obligation is met by acquiring tradable certificates, or renewable energy certificates, to the necessary level. This allows open competition between different forms of renewable generation anywhere in Australia. If retailers fail to purchase sufficient certificates to meet their obligations, they would be financially penalised. To date, compliance with the scheme has been higher than 99%.

The Australian Government renewable energy scheme has been in place since 2001. Revisions have been made based on consultation and economic analysis, retaining the intention and design of the scheme without the destabilisation of radical redesign. The initial target for 2010 was relatively small, requiring only approximately an additional 2% of energy consumed be sourced from renewable energy generation. Investment in renewable energy began faltering in the face of a fully subscribed scheme. A change in the Australian Government in 2007 resulted in the renewable energy policy target being increased to 20% of electricity consumed by 2020 (or 60,000 GWh of renewable energy).

The scheme has occasionally suffered because of other renewable energy policy initiatives both by the Australian Government and State Governments. In 2009, the renewable energy target was split to provide separate targets for renewable energy from small generation systems (especially rooftop photovoltaic solar) and that from large renewable generators such as wind farms.

The State Government also established ambitious renewable energy targets during this period and, while they provided no financial incentive, they clarified state government support for development.

Conclusion

Overall, South Australia has benefited from an electricity market and renewable energy incentives that have focused on creating economic drivers rather than selective incentives or selecting individual projects or technologies. National renewable energy targets, for which the incentive price is determined by its own market, have provided a technology-neutral subsidy that operates separately from the electricity market, and effectively delivered economically appropriate renewable energy generation. The location of these projects has been strongly influenced by regional and locational pricing signals from the real time electricity market dispatch and by the proponent bearing the cost of connection. By designing technical standards and performance obligations that are focused on the connection point rather than on the generating equipment, this has, at a minor increase in the cost of development, maintained a reliable and secure network that can support ongoing renewable energy development.

Institutionalized market planning, based on socialised and accepted technology performance and cost data, provides unbiased information to all market participants. Generators gain insight into the relative potential of their chosen technology over the longer term; transmission companies obtain significant information about network constraints and the relative merit of their removal; regulators and other non-market bodies can clearly see potential market growth and reaction and benchmark the overall efficiency of their policies.

References

[AEMO] Australian Energy Market Operator. (2010). "An Introduction to Australia's National Electricity Market." Melbourne: AEMO. Accessed April 17, 2012: http://www.aemo.com.au/ corporate/0000-0262.pdf..

AEMO. (2011). "2011 Electricity Statement of Opportunities." Accessed April 17, 2012: http://www.aemo.com.au/planning/esoo2011.html..

AEMO. (not dated). "Electricity Market." Accessed April 17, 2012: http://www.aemc.gov.au/ Electricity/Electricity-Market.html.

Australian Bureau of Statistics. (2011). "3101.0 - Australian Demographic Statistics, Jun 2011." Accessed April 17, 2012: http://www.abs.gov.au /AUSSTATS/abs@.nsf/ allprimarymainfeatures/C521664B22CEE131 CA 2579CF000F9AF4?opendocument.

[ESIPC] Electricity Supply Industry Planning Council. (2005). Planning Council Wind Report to ESCOS. http://www.escosa.sa.gov.au /library/050429-PlanningCouncilWindReport-ESPIC.pdf.

APPENDIX C. CASE STUDY: DENMARK

Author: Hugo Chandler, New Resource Partners

Introduction

A key challenge for Denmark will be to attain its very ambitious target for wind energy: 50% of electricity production in 2020, up from 28% in 2011. Denmark is well located to achieve this level of ambition: the country is strongly connected to Germany and Scandinavian countries, which absorb the wind energy that spills over its borders. The share of wind energy over this larger area is much smaller than when the country is viewed in isolation.

New transmission is less of a challenge in Denmark that it is in Germany, for example. Danish interconnections already amount to approximately 80% of the country's peak power demand, and additional lines are under development. Rather, the challenge will be to find the right balance between addressing the wind power integration challenges with increased reliance upon power trading with neighbors that also have major wind energy ambitions, compared to introducing solutions within the Danish borders.

Germany's wind industry association believes an additional 25 GW could be installed on land and at sea by 2020, on top of the 29 GW today (GWEC n.d.). ENTSO-E estimates that in the Nordic region as a whole, meanwhile, wind capacity could rise to approximately 15–20 GW in the same year (ENTSO-E 2010), at which point less than half of Nordic wind capacity would be located within Danish borders.

Output throughout this northern region is likely to be highly correlated. This means that competition for flexible resources such as Norwegian hydropower, to balance these largely wind power ambitions, is going to increase. Denmark may need to increase its domestic flexibility.

Focus of Case Study

The second section of this case study sets the scene for discussion of past and ongoing challenges in the system integration of variable RE. In Denmark, this equates to wind power. Solar PV and wave power deployment are still negligible.[25] This section describes the status of wind power deployment in Denmark and summarizes policy development. Finally, it highlights best estimates of balancing costs in the literature.

The third section of this appendix addresses the key areas wherein action has been taken to facilitate the deployment of wind power. It is divided as follows:

1. Integrated planning of transmission and variable RE deployment
2. Building public support for new transmission
3. Increasing the flexibility of the Danish power system
4. Making markets more amenable to variable RE
5. Advanced system operation and grid code development

The fourth section addresses the lessons that might be learned from the Danish case by other countries with similar system attributes; and the final section presents key conclusions regarding Danish experiences with wind power.

The term "flexibility" will be used a number of times in this appendix. Briefly, it is used to encompass all aspects of a power system that are complementary to variable (sometimes referred to as intermittent) production of electricity. Examples include controllable power plants (e.g., hydropower, gas, coal plants), energy storage (such as pumped hydropower), demand-side management and response, physical trade of electricity with neighbors, power markets that both enable and stimulate more liquid trading, and more dynamic operation of the power system.

Context

The Danish Wind Market

The share of wind energy in electricity production grew vigorously until 2004, when policy support tailed off with a change of government. It recovered in 2009 with the return of stronger political support, to reach 28% in 2011.[26] Penetration is higher in the west of Denmark, amounting to 30% of annual power consumption in 2010, compared to 22% over the country as a whole.

Policy Support

Government support for wind energy in Denmark has a long history. One of the earliest important events was the establishment of a wind energy centre at Risoe National Laboratory, which pioneered the first wind speed atlases. In 1990, a feed-in tariff (FIT) was set for wind power plants, which was the primary driver for the industry. Also in 1990, the Third Danish Energy Plan set up the first national wind power target: 1,300 MW by 2000. In 1993, a new Ministry for Environment and Energy was set up, linking the two portfolios closely. In 2007, a new Ministry for Climate and Energy was created.

On January 1, 2009, the Promotion of Renewable Energy Act (*Lov om fremme af vedvarende energi*) passed into law. It rebooted the installation of wind power with a strong feed-in premium paid to producers on top of the market price. The premium amounted to $44/MWh payable for 22,000 full load hours in 2009 (Energinet 2009a).

Calls for tender are used to set FITs for offshore wind farms, winners being decided based on the lowest tariff required (Energy Policy 2010, Energinet 2009a). This applies for a fixed period, after which the wind farm remains alone in the market place. For example, the *Rødsand Two* offshore wind farm receives in total (premium plus market price) $0.11 per kWh. The premium applies to electricity production amounting to 10 TWh, for a maximum of 20 years (IEA/IRENA 2011).

February 2011 saw the publication of *Energy Strategy 2050*, which resulted from the findings of the Danish Commission on Climate Change Policy. The strategy aims to increase the share of wind and biomass energy to 33% of all energy in 2020, up from 19% in 2009. In terms of electricity, wind power is to contribute 40%. This already ambitious target was increased to 50% by the new government that took power in Denmark in October 2011 (DWEA).

Costs of Wind Power Integration

Integration costs can be broken down into three categories: those that relate to transmission extension and reinforcement (not including the cost of linking to the grid); those incurred in the balancing of increased volatility in the power system; and those that may be incurred to maintain the adequacy of the power system (i.e., its ability to cover peak demand).

In 2008, an Electricity Infrastructure Committee, including Energinet, estimated transmission needs up to 2030, including those resulting from wind power. From the broad range of technology choices it investigated, with different mixes of overhead lines and underground (more expensive) cables, the Danish Government selected the option with a cost attributable to wind power of $354/kWh, for 3.5 gigawatts of additional wind power.

The Greennet study (see Holttinen et al. 2009) found that wind power balancing costs in Denmark at market share of 28% amounted to approximately $1.31–$2.62/MWh.[27] These values are supported by real world experiences in West Denmark where costs have amounted to $1.8 - $3.4/MWh at a 24% penetration (Holttinen et al. 2009).

Key Factors with Bearing on Variable RE

This section addresses challenges, actors, their actions, and outstanding issues related to integrating wind power into the Danish power system.

1. Integrated Planning of Transmission and Variable RE Deployment

The high-voltage transmission grid in Denmark is fully owned by a public company, Energinet, under the Ministry of Climate and Energy. Consequently, reinforcement and expansion of it, as well as international link-ups, are driven directly by the strategic energy policy of the government. The value of Energinet as a driver of effective wind power integration and management in Denmark cannot be overstated. Indeed, in and of itself, Energinet is a principal reason the grid can be said to be limited barrier to wind power.

For example, in the case of major power projects, including larger wind farms, Energinet will usually commence front-end work (such as design and consenting) ahead of the RE project consent. This practice is not regulated as such; the TSO can use its discretion as to timing. Although it does not actually commence construction until the power plant is consented, this early action is likely to reduce the length of the integration process, which is important as reinforcement work can take two to three years to complete (Eclareon 2011).

In March 2009, Energinet completed a Cable Action Plan in collaboration with regional transmission companies, which details the upgrading of the entire high voltage grid. This amounts to approximately 3,200 kilometers (1,998 miles) of line to be replaced with about 2,900 kilometers (1,802 miles) of new cables (Energinet 2009b). The action plan addresses not only wind power; it is a coherent plan for the needs of the system as a whole. Work that benefits wind power deployment also substantially benefits the wider system, and vice versa.

East and West Denmark are part of distinct grid systems. The Jyland Peninsula is part of the continental European system, while East Denmark–comprised of the islands–is part of the Nordic system. While both have long interacted through the Nordic power market, they were finally electrically connected in 2010 via the "Great Belt" direct current link, providing additional management options for surplus wind power in the west.

International Transmission Efforts

Significant efforts are underway to electrically link Denmark more closely to adjacent and nearby markets, and some links have already been made. These links help spread inexpensive wind power production more widely, increasing its value, while decreasing power prices in the areas to which it flows. They also enable surpluses to be stored (e.g., in Norwegian pumped hydropower facilities). In an average year, imports and exports of electricity to and from Denmark amount to the equivalent of nearly one third of total consumption.

Additional interconnector links to Norway the Netherlands, both of them marine cables, are being developed. *Skagerrak 4* is an additional 600-MW high-voltage direct current (HVDC) connection with Norway across the stretch of sea with that name, scheduled for operation in 2014. The final investment decision regarding the *Cobra* 700-MW HVDC link between Jyland and the Netherlands is planned to be made in the end of 2014.

Coordinated Planning and Connection of Variable RE Plants

Denmark and Germany are planning an important technological innovation that may prove to be the first step taken toward an international, truly offshore grid.[28] It may be an efficient way to share the output of large wind farm clusters among sponsor countries, while also increasing the capability of power to flow between the latter when the wind is not blowing. The wind farm cluster will be located at Kriegers Flak in the Baltic Sea, between the sponsor countries.

Grid Connection

Most land-based wind power in Denmark is connected to the distribution grid. Under the Electricity Supply Act, the distribution system operator is obliged to connect any new power plant meeting pre-established technical requirements. The Act also gives priority to RE plants. However, the precise connection procedure is not codified in law, which can lead to delays.

The distribution system operator is obliged to reinforce the grid if necessary, in consultation with Energinet, and the distribution system operator pays. The cost is passed on to the consumer through a levy, the Public Service Obligation. This is significant as the connection cost usually represents less than 10% of the total integration cost (RETD 2008). Wind energy, specifically, gets further financial help under the Promotion of Renewable Energy Act, which specifies that the developer and the distribution system operator share the cost of connection. RE plants are also exempt from charges on the use of the transmission system (Eclareon 2011).

Energitilsynet, the energy regulator, must be consulted if the development of the grid at the 100- kV level or higher is concerned; it must balance RE deployment ambitions with least cost to the consumer. The regulator is not obliged to set a deadline for this procedure, but again, the delays encountered do not appear to be significant. Indeed, a recent study (Wind Barriers 2010) found that the average time required for a grid connection permit in Denmark was an order of magnitude less than in the wider European Union.[29]

Developers in offshore development areas bear no integration costs outside the wind farm itself, the boundary of which is usually measured as one rotor's width from the turbines. The link to the grid—including the transformer station, cable to shore, and any reinforcement required of the land grid—is considered integral work to the grid, and is therefore met by Energinet and passed on to the consumer (Van Hulle et al. 2009).

Variable RE Planning

The more than 5,000 onshore wind turbines installed in Denmark to date have resulted from no centralized spatial planning process. Local decision-making has driven their distribution, quite evenly, over Denmark, although concentrations of turbines are higher in the west (Jyland) and coastal regions.

Since 2010, a Wind Turbine Secretariat has assisted municipalities in the entire planning process. *Energy Strategy 2050* suggests combining this decentralized approach with new, more holistic, planning tools. For example, the strategy includes a proposal to locate turbines nearer roads and railways; however, doing so is currently hindered by distance requirements. The strategy targets some 1,800 MW of new onshore wind plants and estimates that some 500 MW of this additional capacity will be brought about by this improved planning approach (Danish Government 2010).

For specific areas identified through a screening process, the government steers the location of new offshore farms through public calls for tender. The youth of offshore wind technology is such that any development outside zones designated by government calls for tender would be prohibitively expensive, as developer would have to bear any integration costs.

Specific targets for additional offshore wind power include two calls to develop 600-MW wind farms at Kriegers Flak in the Baltic Sea and at Horns Rev in the North Sea, and an additional 400 MW of smaller, near-shore installations. And, plans may be afoot to open new offshore development zones to meet the recently announced 50% wind target for 2020 (RECHARGE 2011).

No action has been identified that was specifically taken by Danish actors to diversify location of wind power in order to achieve geographic smoothing of aggregated output. This is perhaps because Denmark is too small for much gain to be made, and plants are already widely spread (see Figure C-2).

2. Building Public Support for New Transmission

Very strong public support exists for RE in Denmark, but it does not necessarily stretch to the additional transmission infrastructure that wind power implies. Moreover, as in most Organisation for Economic Co-operation and Development countries, grid infrastructure in Denmark is aging and in need of routine upgrading.

Denmark plans to take radical action to remove public objections—mainly aesthetic objections— to new transmission needs. In February 2008, political agreement was reached on the need for cabling (i.e., placing lines underground), with the selection of one of six options for grid development prepared by an Electricity Infrastructure Committee. Energinet and its regional partners subsequently prepared a Cable Action Plan, and presented it to the ministry in March 2009. The plan includes the undergrounding of the entire 132-kV to 150-kV grid by 2030, and all new 400-kV lines by 2030. The committee also suggested that the entire distribution network (6– 60 kV) might be buried in this timeframe. The cost of undergrounding the 132-kV to 150-kV grid was estimated to be $2.0 billion, while the cost of burying the distribution grid was estimated to range from $1.7 billion to $2.0 billion (Danish Electricity Infrastructure Committee 2008).

3. Increasing the Flexibility of the Danish Power System

Energinet sees electricity as the principal energy carrier in the future, increasingly covering the energy needs of the heating and transport sectors (Energinet 2012a). If this were indeed the case, increased electric heat and transport would represent very important temporal buffers against the variability of wind and solar PV.

Combined Heat and Power Production

All combined heat and power (CHP) plants must participate in the spot power market (Power & Energy 2009). Consequently, their power output reflects the electricity price to a greater extent than in the heat-load-following mode that is more usual for CHP plants. This is possible because most CHP plants in Denmark produce low-temperature steam for district heating, representing a considerable heat store (Frontier Economics 2010), rather than high temperature heat for industrial processes, which is harder to store.

Given that nearly all of Denmark's thermal electricity production—more than 55% of all electricity production (Danish Energy Agency 2011)—is from CHP plants, this market participation represents considerable additional flexibility. In addition, approximately one third of small CHP plants are active in the regulating power market, contributing to ancillary system services (Power & Energy 2009).

In 2008, the tax levied on the use of electricity for heat production at CHP plants was reduced. As a result, when the price of electricity is low, or negative—as happens when wind output is at maximum—using this (wind) electricity to produce heat, instead of burning oil or gas, is cost effective. So, electricity production from fossil fuels can be reduced, providing downward flexibility when it is most needed to accommodate wind power (Power & Energy 2009, Energinet 2009a) while in addition, more of the latter finds a local use.

Additional Sources of Domestic Flexibility

Electric cars may serve as a source of system flexibility; if effort is made to correlate their charging times with times of electricity surplus—when wind is at maximum—they can serve as a store. Parked vehicles might also return power to the grid when the wind falls away. Electric cars are tax free in Denmark (Energinet 2009a), and the *Energy Strategy 2050* includes $40 million to support the deployment of charging stations.

Energinet is participating with other Danish and international partners in a project known as EcoGrid EU that will demonstrate the ability of the island of Bornholm to provide itself with 50% renewable electricity, mainly wind power, with active management of the distribution system, electric vehicles, and technology and price incentives to enhance demand side response. The intention is to demonstrate, on a small-scale, approaches that could be scaled up (Energinet 2012b).

The *Energy Strategy 50* also contains the intention to prepare a "smart grid" strategy. The Danish government is seeking an agreement with the distribution companies (of which there are 80), to install intelligent electricity meters when consumers install heat pumps or vehicle recharging stations. The level above which intelligent meters must be installed will be reduced from 100,000 kWh to 50,000 kWh in 2013.[30] All meters replaced after 2015 must be replaced with intelligent meters, to enable greater response from the demand side. To reinforce consumers' incentives to respond to more volatile electricity prices resulting from higher wind share, a range of new initiatives is being developed. These include flexible settlement systems, a data hub (Energinet 2012c), and efforts toward hourly settlement.

4. Making Markets More Amenable to Variable RE

Energinet and the other Nordic TSOs are responsible for developing the Nordic power market design. Thus, the organization partly responsible for overall market efficiency is also the organization with a strategic objective to integrate more variable RE. From an institutional perspective, this is a very complementary arrangement.

An Ever-widening Power Market

In the 1990s, the newly deregulated markets in the Nordic region began to integrate. Cross-border tariffs were removed, and a common power exchange, Nord Pool, was established originally for only Norway and Sweden. All four Nordic national system operators, including the Finland TSO and Energinet, now jointly own Nord Pool. West Denmark joined in 1999, and East Denmark joined in 2000. Text Box C-1 explains briefly how the exchange manages the flow of electricity across national borders.

Text Box C-1. Congestion Management through Market Splitting

A key aspect of this collaboration is the efficient allocation of cross-border transmission capacity (a task known as "congestion management"). In the Nordic market, allocation is implicit in the spot market trade of electricity; traders need not subsequently acquire the right to use cross-border transmission in an "explicit" auction. The exchange does this through market "splitting." This has a similar effect to the Central and West Europe market "coupling" to the south, in that flows across borders are optimized without being explicitly assigned to any particular trade. Variations in price can develop in the different control zones of the Nordic Market,[1] which result from transmission bottlenecks distorting the system price. When this occurs, the system operator—through the day-ahead market—will buy electricity in the cheaper area and sell it in the more expensive area, causing power to flow from the low price area toward the high price area, at the same time reducing the price difference between them.

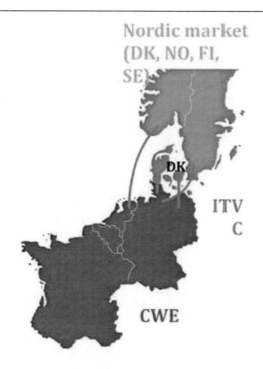

Figure C-1. Collaboration of Nordic and Central and West Europe Markets through the Interim Tight Volume Market Coupling.

The success of Nordic collaboration is reflected in the fact that trade over the area increased by a factor of 3.5 from 1975 to 2006, while demand for electricity only doubled. Approximately 75% of all Scandinavian electricity is traded in Nord Pool, up from 45% in 2006 (VTT 2007). Participation in the exchange is not mandatory, but only through it can agents trade over the entire Nordic area; bilateral trades are limited to the control zone.[31]

Increasing international trade therefore increases the liquidity of the exchange. This not only brings benefit in terms of wider system efficiency and cost reduction; it means that increasingly large amounts of controllable power are available, so that greater amounts of wind power can be accommodated efficiently and securely.

In 2010, another step was taken toward a single European power market, with the establishment of the Interim Tight Volume Market Coupling of the

Nordic and Central and West Europe markets,[32] which enlarged the area over which electricity can be exchanged (Text Box C-2).

The Interim Tight Volume Market Coupling functions on the day-ahead time scale (i.e., intra-day and shorter-term trades remain within the Nordic area) although Germany also participates in the Nordic Intraday market. As its name implies, the Interim Tight Volume Market Coupling is intended to be a step toward deeper integration of the two markets. A permanent version of the coupling is hoped for by the end of 2012 (Energinet 2012a).

Text Box C-2. The Interim Tight Volume Market Coupling

In August 2008, the Electricity Market Coupling Company was founded as a joint venture of Nord Pool Spot, the European Energy Exchange (EEX), Nord Pool Spot AS, the German TSOs 50Hertz Transmission GmbH and Tennet TSO GmbH, and Energinet.

The Electricity Market Coupling Company established a coupling of Danish and German markets in 2008, an improved version of which was launched in 2009. This in turn was developed into the Interim Tight Volume Market Coupling. The Electricity Market Coupling Company encourages optimum electricity flows over the coupling based on the transmission that is available to carry electricity at any given moment, which information is provided by the TSOs; and the trading of electricity for that moment, which information is provided by the exchanges.

Nordic Market Design

Renewable electricity has priority dispatch under the Electricity Supply Act. Although there are no trading rules specifically for variable RE in the Nordic market, it has been encouraged to evolve to make more efficient use of ever-growing amounts of wind power. Indeed the evolution of the market is continuous, with important steps toward flexible operation seen regularly.

The core of the market is a day-ahead platform, Elspot, where the bulk of electricity is traded, from 12 hours up to 36 hours ahead of delivery. This is quite a good timescale for thermal plants, providing plenty of time to ramp production up or down, but it is less suited to fluctuating and less-certain wind power production.

When trading has closed on Elspot, it can continue through the intraday market, Elbas, with continuous trading up to 60 minutes ahead of delivery.[33] Though the volume of power traded on Elbas is very much smaller (approximately 1% of exchange traded power), it is increasing (Ea Energy

Analyses 2011), which reflects a growing share of uncertain wind power production in the market place. This in turn represents a growing challenge to system operators for whom planning of system operation will be increasingly time-constrained.[34]

A regulating power market operates up to 15 minutes before delivery.[35] This is not part of the Nord Pool exchange; offers to provide (or reduce consumption of) power are made directly to Energinet, which passes them on to the Nordic Operational Information System , along with offers from the other TSOs. Because of the Nordic Operational Information System, therefore, providers of short-term flexibility—which includes consumers who can offer to reduce their consumption (Bjørndal 2002)—can be shared over the whole Nordic area. In addition, Energinet is pursuing opportunities to open a Europe-wide balancing market (Energinet 2012a).

Between them, these three markets provide considerable ability for producers, retailers, and consumers to calibrate and recalibrate their trades in the face of the increasing variability and uncertainty resulting from a larger wind power share.

Negative Prices

In 2009, negative prices were permitted for the first time in the day-ahead market, in which previously there had been a floor price of $0/MWh.[36] Negative prices may occur, for example, when wind turbines are at full power but electricity demand is low. At present, they occur for about 20–100 hours per year in Denmark. Even more so than by a zero price, electricity producers are encouraged to reduce their production at such times or be in the perverse situation of having to pay to generate.

The intention is to use the price signal to encourage participants to be more ready to respond to market needs. This includes wind power producers, who are also encouraged to reduce their production at a certain point. Additionally, large consumers that pay the spot price will be encouraged by negative prices to shift consumption to these times if they are able.

5. Advanced System Operation and Grid Code Development

Energinet is the sole operator for both parts of Denmark, following the statutory merger of the smaller TSOs, ELTRA and ELKRAFT, in December 2004. This is likely to have increased the coherency and efficiency of management of the whole. The challenge of managing the Danish power system has grown enormously over the last 30 years; Figure C-2 shows the number of power plants in 1980 compared to 2010. Nevertheless, Denmark has not suffered any major system contingencies resulting from wind power.

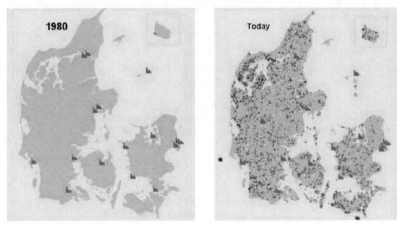

Source: Energinet (2009a).

Figure C-2. Number of power plants in Denmark, 1980 and 2010.

Energinet has greatly enhanced its energy management system to monitor real-time performance of the power system.[37] This includes real-time estimates of wind power to be fed into the grid. This is an important; because most wind power in Denmark is connected at the distribution level, which is passively managed, it is not very visible to the system operator (Jones 2011).

Energinet uses multiple, advanced forecast tools, resulting from rolling innovation of wind output forecasting tools in Denmark over 20 years. Energinet has played a central role in this innovation, which has included Danish power industry and universities also. Forecasts are used to plan system operation, day-ahead (cross-border) congestion management, the commitment and economic dispatch of controllable power plants, contingency analysis, the assessment of grid transfer capacity, and the need for regulating power.

Energinet developed the *Drift Planlaegnings* (Operational Planning) System tool for integrating forecasts of wind and CHP output into system planning up to two hours ahead of the time of operation. To better manage more wind power, Energinet is in the process of pushing this horizon up to five minutes (Jones 2011). The tool also provides the operator with a view of power flows between the Danish system and those of Germany, Norway, and Sweden.

Energinet requires all power plants of greater than 10-MW capacity to provide automatic updates every five minutes of their power production. Combined with its own estimates of wind power and small CHP output, this allows Energinet to track the output of the complete range of power plant on the Danish system (Jones 2011).

The increasing share of wind power, resulting in a greater but less predictable need for short-term flexible resources, has been a principal driver of closer cooperation among Nordic TSOs, which seek to better manage and share their flexible resources. In 2004, the Nordic Operational Information System (NOIS) was established; it is based at Energinet. NOIS provides common understanding of the status of interconnections, and usable operating reserves over the Nordic area.

Higher-Resolution System Operation

The Nordic day-ahead market is dispatched in hourly blocks, a practice that is less well suited to variable output producers than, for example, fifteen minute blocks are. The latter allow greater fine-tuning of trades to account for variability and uncertainty. Smaller individual blocks of power fit better with the more volatile net load,[38] reducing scheduling inefficiencies encountered when longer blocks are used. Though this increases complexity, it may also result in a reduced need for short-term balancing resources.

In the long-term, ENTSO-E considers that markets also will need to move to fifteen minutes periods, in order to provide greater incentive to balancing resources to respond to system needs (ENTSO-E 2010).

Incentivizing Variable RE Forecast Accuracy

Before 2003, wind turbine owners had no responsibility to produce according to their forecasts. Then, policy changed to require wind turbines to pay the costs of being out of balance (i.e., for producing more or less than what is forecast and sold in the day-ahead market). To compensate for this, wind warm operators can claim a refund of $3.9/MWh (IEA 2012) on top of the feed-in premium, under the Promotion of Renewable Energy Act. Thus, the operator of a wind farm has incentive to minimize the divergence from forecast output, while being compensated to some extent, the cost of which is spread among all consumers.

Grid Codes

The Nordel Grid Code, launched in 2007, applies across the whole Nordic area. In addition, Denmark has its own requirements. The earliest version was applied in 1999, and it required wind turbines to be able to *disconnect* from the grid during abnormal voltage and frequency events. In 1999, a new code was published that applied to wind farms on the high voltage grid only, mainly targeting offshore wind farms.

This new code required the opposite of wind turbines: that they remain connected and delivering electricity during the same kinds of events; in other words, from then on, wind farms were expected to support the grid in case of fault, not to disconnect at the first sign of trouble. It was a world first, with several other requirements of wind farm owners, including that new wind farms at this voltage level should be controllable remotely so that they can be curtailed if necessary (Energinet 2009a).

Given that 90% of wind turbines are connected at the medium voltage (60-kV) level and below, similar grid codes now also apply at that level. The grid code is revised to keep pace with development of wind turbine technology, as wind technology becomes better able to provide other such ancillary services to support the power system.

Curtailment

Curtailment rules prefer RE sources, meaning that only after fossil-powered plants have been reduced to minimum can curtailment of wind power be done. Curtailment is required relatively rarely, when storm conditions occur during cold periods (driving high CHP output) with low electricity demand. It applies to turbines that can be controlled centrally. Until 2007, curtailment was not compensated (Van Hulle et al. 2009), but Energinet now pays compensation to land-based owners.

However, the Promotion of Renewable Energy Act took into consideration that the need to curtail may increase, particularly with the advent of very large wind farms offshore. Under the Act, the output of certain offshore wind farms may be curtailed if required for security reasons.[39] This is only compensated by Energinet if the decision is taken as a result of a sudden fault. If notice is given day-ahead, no compensation is paid (Energinet 2009a).

Offshore wind farms may be required to control their power output in several ways, some of which are captured in Figure C-3. These highlight the capabilities of modern wind turbines. In the first case, a simple ceiling can be placed on the output. In the second, it is ordered to reduce to a fixed level for a certain period, while in the third, the rate at which output increases (i.e., according to the wind speed) can be artificially smoothed. In the fourth graph, actual output has been reduced by a fixed amount over time, with the result that during that period, a certain amount of additional power is available if required by the system operator for balancing; in effect, the wind turbine has itself become a source not only of downwards flexibility, but also upwards flexibility.

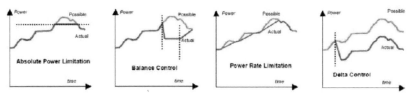

Source: ENTSO-E 2010.

Figure C-3. Active power control abilities of Danish offshore wind farms.

Potential for Learning in Other Countries

All power systems are different, and the great success in terms of wind power deployment witnessed in Denmark does not mean all countries could easily have the same wind power share.

Denmark is a small system, heavily interconnected with both Scandinavian neighbors in the Nordic power market and Germany to the south, with a transfer capacity equal to approximately 80% of its peak demand. In other words, surpluses and deficits of power production resulting from a large variable RE share can relatively easily be compensated for. Other systems are likely to have a far smaller potential to trade, relative to their size.

Denmark's experiences with increasing the flexibility of CHP power plants could be exported to other countries where cogeneration of heat and power can be decoupled to a considerable extent, as is possible in Denmark because of the high proportion of CHP plants linked to district heating systems.

Danish system operational experience, particularly in the design and use of forecasting tools, represents best practice that should be embraced by countries seeking a large share of variable RE. Similarly, the market practices, transparency, and collaboration of Denmark with its neighbors provide an example to other countries of how to share flexible, controllable balancing resources in order to manage wind power.

Conclusion

Denmark has a strong transmission grid that is monitored and maintained closely by a single owner, Energinet, which is managed directly by the Ministry of Climate and Energy. Consequently, Energinet's efforts very closely align with strategic government policy, including policies regarding wind power deployment. In 2009, Energinet developed a holistic plan for the complete overhaul of the high-voltage grid to enable Denmark to meet the 50% wind power by 2020 target. The plan, when implemented, will also address the single greatest non-economic barrier to wind power, public antipathy to new grid infrastructure, by undergrounding much of the high-voltage grid.

Grid connection is, at worst, only a slight barrier to variable RE deployment in Denmark, as grid companies have taken swift action to expedite connections. Deployment of wind turbines on land has been managed in a decentralized manner, but this is changing to best use remaining sites and reach the 50% wind power target by 2020. Tender processes have steered offshore wind farms into pre-screened zones. Denmark is also, with Germany, planning the first offshore grid in the Baltic Sea, which will connect a cluster of offshore wind farms to the two countries.

Denmark's embedded position in the Nordic power market is her greatest wind integration ally. Consequently, Denmark is taking important steps to reinforce this position, with a new cable to Norway and a new link to the Netherlands to the south. Trading of controllable power resources to complement variable RE output can be traded across the entire Nordic market. This ability to share resources extends right up to the time of operation, through a shared regulating power market.

Wind producers are exposed to the market; they sell their power into it, and if they are out of balance at settlement time, they incur costs. This encourages them to keep as close as possible to their output forecast, and they are compensated to some extent with flat payments on top of the feed-in premium. On land, wind curtailment is a measure of last resort, and is compensated. New offshore wind farms may receive notice to control their production in several ways.

Domestic flexibility is being enhanced with more flexible operation of—very common—CHP plants, plans for electric vehicles, and large-scale deployment of smart meters. Using negative prices in the spot markets adds incentive to controllable and variable power plants alike, to give more consideration to the needs of the market. Offshore wind turbines can be not only curtailed but also operated below maximum so that they can provide up regulation[40] if needed.

System operation is overseen by a single TSO, Energinet, and it has become significantly more complex over the last 20 years. Energinet and the Danish power producers have pioneered practices to manage this complexity. Denmark was the cradle of output forecasting techniques, and it continues to lead the development of system operation planning tools.

References

Bjoerndal, M.; Joernsten, K.; Pignon, V. (2002). "Congestion Management in the Nordic Power Market – Counter Purchases and Zonal Pricing."

Danish Electricity Infrastructure Committee (*Elinfrastrukturudvalget*). (2008). "Technical Report on the Future Expansion and Undergrounding of the Electricity Transmission Grid: Summary." https://selvbetjening.preprod. energinet.dk/NR/rdonlyres/CC966C3A-FE78-41D8-9DC7-6B455210B 502/0/TechnicalReportSummary.pdf. Accessed April 16, 2012.

Danish Energy Agency. (2011). Energy Statistics 2010. http://www.ens.dk/en-US/Info/FactsAndFigures/Energy_statistics_and_indicators/Annual%20St atistics/Documents/En ergy%20in%20Denmark%202010.pdf. Accessed April 16, 2012.

Danish Government. (2010). "Energy Strategy 2050 – From Coal, Oil and Gas to Green Energy" Copenhagen, Denmark.

[DWEA] Danish Wind Energy Association. (October 12, 2011). "Danish Government: '50 percent of power consumption from wind power in 2020'." Accessed April 16, 2012: http://www.windpower.org/en/news /news.html#720.

Ea Energy Analyses. (2007). "50% Wind Power in Denmark in 2025." Copenhagen: Ea Energy Analyses.

Ea Energy Analyses. (2011). "The Existing Nordic Regulating Power Market." Copenhagen: Ea Energy Analyses.

Eclareon. (2011). "Integration of Electricity from Renewables to the Electricity Grid and to the Electricity Market: Draft Final National Report. Denmark.: Berlin: Eclareon.

Energinet. (2009a). "Wind Power to Combat Climate Change: How to Integrate Wind Energy into the Power System" Fredericia, Denmark: Energinet.

Energinet. (2009b) "Cable Action Plan, 132 - 150 kV Grids." Fredericia, Denmark. http:// energinet.dk/SiteCollectionDocuments/Engelske%20 dokumenter/Om%20os/Cable%20Action%20Plan%20-%202008-2009. pdf.

Energinet. (2012a). "System Plan 2011" Copenhagen: Energinet.

Energinet. (2012b). "EcoGrid EU." Accessed April 12, 2012: http://www.energinet.dk/EN/FORSKNING/EcoGrid-EU/Sider/EU-Eco Grid-net.aspx.

Energinet. (2012c). "DataHub." Accessed April 16: 2012: http://energinet.dk /EN/El/Datahub/ Sider/DataHub.aspx.

ENTSO-E. (2010). "Impacts of Increased Amounts of Renewable Energy on Nordic Power System Operation." Brussels: ENTSO-E.

Frontier Economics. (2010). "Study on Flexibility in the Dutch and NW European Power Market in 2020." Prepared for Energinet.

[GWEC] Global Wind Energy Council. (no date) "Germany: Total Installed Capacity. Accessed April 16, 2012: http://www.gwec.net/index. php?id=129.

Holttinen, H; Meibom, P.; Orths, A.; van Hulle, F.; Lange, B.; O'Malley, M.; Pierik, J.; Ummels, B.; Tande, J.O.; Estanqueiro, A.; Matos, M.; Gomez, E.; Söder, L.; Strbac, G.; Anser Shakoor; Ricardo, J.; Smith. J.C.; Milligan, M.; Ela, E. (2009). "Design and Operation of Power Systems with Large Amounts of Wind Power. Final report, IEA WIND Task 25, Phase One 2006–2008." Vuorimiehentie, Finland: VTT. http://www.vtt.fi/inf/pdf/tiedotteet/2009/T2493.pdf

IEA/IRENA. (2012). IEA and IRENA Policies and Measures Database. http://www.iea.org/ teieaxtbase/pm/?mode=re&action=detail&id=4425. Accessed February 3, 2012.

Jones, L.E. (2011). "Strategies and Decision Support Systems for Integrating Variable Renewable Energy Resources in Control Centers: Global Best Practices, Examples of Excellence, and Lessons Learned." Prepared by Alstom Grid for the U.S. Department of Energy. Washington, DC: Alstom Grid.

Klinge Jacobsen, H.; Zvingilaite, E. (2010). "Reducing the Market Impact of Large Shares of Intermittent Energy in Denmark." *Energy Policy.*

RECHARGE. (October 21, 2011). "Update: Danes Eye New Offshore Zones to Hit Higher Wind Target." Accessed April 16, 2012: http://www. rechargenews.com/energy/wind/ article284551.ece.

[RETD] Renewable Energy Technology Deployment. (2008). "Innovative Electricity Markets to Incorporate Variable Production," Prepared by IPA

Energy and Water Economics for the IEA Renewable Energy Technology Deployment Implementing Agreement.

Van Hulle, F.; Tande, J.O.; Uhlen, K; Warland, L. ; Korpås, M.; Meibom, P.; Sørensen, P.; Morthorst, P.E.; Cutululis, N.; Giebel, G.; Larsen, H.; Woyte. A.; Dooms, G; Mali, P-A; Delwart, A.; Verheij, F.; Kleinschmidt, C.; Moldovan, N.; Holttinen, H.; Lemström, B.; UskiJoutsenvuo, S.; Gardner, P.; van der Toorn, G.; McLean, J.; Cox, S.; Purchala, K.; Wagemans, S.; Tiedemann, A.; Kreutzkamp, P.; Srikandam, C.; Völker, J. (2009). "Integrating Wind: Developing Europe's Power Market for the Large-Scale Integration of Wind Power, Results of the TradeWind Project, coordinated by the European Wind Energy Association, Brussels. TradeWind.

[VTT] Technical Research Centre of Finland. (2007). "Optimising the Market Integration of Electricity from Renewables." Holttinen, H.; Korenoff, G.; Lemström, B. Helsinki, Finland: VTT.

APPENDIX D. CASE STUDY: GERMANY

Author: Hugo Chandler, New Resource Partners

Introduction

Challenges of Integrating Variable RE

Germany must manage very large flows of wind energy into and around its grid area. Until recently, with the scaling up of solar photovoltaic power plants (PV) in the south of the country, almost all variable renewable energy (RE) generation (i.e., wind power) has been in the middle and north if the country. The lack of balance between rural areas with high wind energy shares and principal consumption areas all over Germany has led to transmission congestion between these different areas.

The challenge is likely to be compounded by growing flows of variable electricity from outside Germany's borders. Germany's immediate neighbor to the north is Denmark, which targets 50% wind power. Moreover, wind penetration is likely to be highest in the Jutland Peninsula, which is part of the same power system as Germany (i.e., the synchronous grid of continental Europe). Instantaneous shares in Jutland can already rise above 100% today.

Grid congestion in the border region during times of high wind is likely to increase without reinforcement.

In addition, flows of electricity from Germany to and through Eastern neighbors are already challenging, to the extent that eastern neighbors are considering remedial measures.

Finally, fast-growing, distributed solar photovoltaic (PV) installations in the south of the country will increase the complexity of the system operation task, particularly because the distribution grid is managed passively.

Focus of Case Study

The second section of this case study sets the scene for discussion of past and on-going challenges in the system integration of variable RE. It describes the status and likely development of the market for wind and solar PV, and it summarizes the history of policy development in the country. Finally, it highlights best estimates of system integration costs in the literature.

The third section addresses the key areas wherein action has been taken to remedy barriers to the deployment of variable RE. It is divided as follows:

1. Coordinating planning initiatives
2. Building public support for new transmission
3. Increasing the flexibility of the power system
4. Making markets more amenable to variable RE
5. Promoting diversification of location
6. Advanced system operation and grid code development

The fourth section addresses lessons that might be learned from the German case by other countries with similar system attributes—such as characteristics related to the grid, RE resources, and flexibility (see below)—and, the last section presents key conclusions regarding the German experience of variable RE.

"Flexibility" will be referred to a number of times in this appendix. Briefly, the term is used here to encompass all aspects of a power system that complement variable generation. Examples include dispatchable power plants (i.e., those than can be turned up or down to greater or lesser extents, energy storage, demand-side management and response, trade, dynamic operation of the power system that enables it to better manage uncertainty in variable RE output forecasts, and markets that both enable and stimulate more iterative trading of electricity).

Context of Integration

Wind and Solar PV Penetration of Electricity Supply

Since the implementation of the first major renewables legislation in 1991, renewable energy in Germany has diversified away from a purely hydropower base to include significant quantities of wind, biomass and, more recently, solar PV generation (Figure 1).

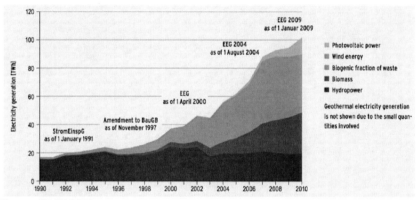

Source: BMU 2011a.

Figure D-1. Development of electricity generation from RE in Germany since 1990.

Table 1 shows the average annual share of wind power in total electricity generation increasing from 1% in 1999 to 6% in 2010. Solar PV, from a much later start, reached nearly 2% in 2010. While these figures seem still quite modest, instantaneous shares can be very challenging.

Table 2 shows the maximum ratio of solar PV and wind power output to power demand in Germany as a whole and in the four TSO control areas into which it is divided (See Figure 2). Perhaps surprisingly, given the modest annual figures above, penetration reached over 60% on Sunday May 8 at 1:00 p.m., when demand dropped to a low on a quiet, sunny afternoon. At the same time, in the area managed by TenneT, which stretches from the north to the south of the country—picking up power both in the windy north and in the sunny south—penetration reached 160% of the entire demand of the area. Eastern Germany saw similarly little activity at 6:00 a.m. on January 1, 2011, and the system operator (50Hertz) had to manage wind output amounting to 124% of the area's demand.

Table D-1. Shares of Wind and Solar PV Generation in Total Electricity Generation in Germany since 1999 (GWh)

Year	Wind (GWh)	Solar PV (GWh)	Total Electricity Generation	Wind Share (%)	Solar PV Share (%)
1999	5,528	42	556,300	1.0	0.0
2000	7,550	64	576,543	1.3	0.0
2001	10,509	76	586,406	1.8	0.0
2002	15,786	162	586,694	2.7	0.0
2003	18,713	313	606,719	3.1	0.1
2004	25,509	556	615,287	4.1	0.1
2005	27,229	1,282	620,574	4.4	0.2
2006	30,710	2,220	636,761	4.8	0.3
2007	39,713	3,075	637,100	6.2	0.5
2008	40,574	4,420	637,232	6.4	0.7
2009	38,639	6,583	592,464	6.5	1.1
2010	37,793	11,683	627,918	6.0	1.9

Sources: BMU 2011a, Eurostat 2011.

Table D-2. Maximum Ratio of Wind and Solar PV to Load, by TSO, in Germany in 2011

	50Hertz	Amprion	EnBW TNG	TenneT	Germany
Date and hour	01/01/2011	07/09/2011	25/04/2011	08/05/2011	08/05/2011
	06:00	14:00	13:00	13:00	13:00
Wind and solar PV (MW)	6,383	3,921	1,847	11,021	20,697
Wind	6,383	2,404	54	5,034	8,070
Solar PV	0	1,517	1,793	5,987	12,627
Load (MW)	5,145	11,082	4,617	6,876	34,435
Maximum ratio of variable RE to load (%)	124%	35%	40%	160%	60%

Source: Prognos 2011.

Policy Support

The German government strongly supports renewables deployment and is a main driver behind variable RE deployment. It contains strongly supportive cooperating ministries and agencies.

Chief among these, from the point of view of RE, are the Federal Ministry for the Environment, Nature Conservation and Nuclear Safety (BMU), which was established after the Chernobyl nuclear reactor event of 1986; and the

Federal Ministry of Economics and Technology, which is responsible for legislation over the wider electricity industry, including the grid. The German Energy Agency (DENA), half-owned by the Federal Republic of Germany and half by financial institutions,[41] provides expertise with respect to renewable energy, energy efficiency, and intelligent energy systems, while the Federal Networks Regulator (BNetzA), under the Federal Ministry of Economics and Technology, regulates energy supply networks.

Ahead of most other countries, in 1989, the BMU introduced the Market Stimulation Programme for installation of 250 MW of wind power. Among other incentives, this guaranteed a fixed payment per kWh of electricity produced. It closed in 1995, by which time the BMU had introduced the Federal Electricity Feed-In Law (*Stromeinspeisungsgesetz*) in 1991. It also featured production-based support, and it was replaced by the 2000 Renewable Energy Sources Act (EEG, *Erneuerbare Energien Gesetz*).[42]

An important feature of the EEG is comprehensive monitoring of results, on which basis modifications to the law are made regularly (approximately every three years). The EEG is a good example of comprehensive legislation. The feed-in tariff (FIT) is valid for 20 years in most cases,[43] stepped over time to help manage upfront investment costs, banded according to the maturity of the technology (i.e., more support for younger technologies). Later applicants get less support than early-movers do, as FIT rates reduce from year to year (a process known as degression).

Digression is intended to encourage technology cost reduction, but sometimes it can prove difficult to keep pace with the latter, as when solar PV costs more than halved from 2009 to 2011, faster than the rate of digression in the FIT. When this happens, there is a risk of inappropriate return to developers and cost to the consumer.

Digression also accelerates deployment, encouraging developers to move more quickly to take advantage of earlier (higher) tariffs. The cost of the FIT is borne by the consumer via the four system operators who market EEG electricity on behalf of the consumers.

The Market Stimulation Programme has provided grants since 2000. These initially included the power sector, but they are now exclusively for the heat sector. Another important driver is the public bank, *Kreditanstalt fuer Wiederaufbau* (KfW).[44] KfW provides long-term, fixed, low interest investment loans, and loan guarantees, to projects, amounting to some €10bn by 2008 (RETD 2008). Recently KfW announced EUR 5 billion ($6.6 billion) of loan guarantees to offshore wind projects up to 2020 (Platts 2011). In 2010 alone, it provided EUR 11 billion ($15.5 billion) "for the construction of facilities using renewable energies," including heat)[45].

Deployment targets are important to both signaling required investment volume and deploying new grid assets. The BMU has developed and regularly updated targets, the most recent of which are part of the September 2010 Energy Concept and the EEG (2012). There is no specific target for particular technologies. The Energy Concept targets are as follows:

- 18% of energy consumption by 2020; 30% by 2030; 45% by 2040; and 60% by 2050
- 35% electricity consumption by 2020; 50% by 2030; 65% by 2040; and 80% by 2050.

These targets are highly ambitious. The Energy Concept was updated in summer 2011 following the government's decision to phase out nuclear power by 2022, which represented approximately 23% of German capacity in 2011, after the events at the Fukushima Daiichi nuclear plant in Japan in March 2011. The change of policy resulted in additional promotion of renewable electricity as well as conventional options such as coal power.

Costs of Balancing Wind Power

Integration costs can be broken down into three categories: those that relate to transmission extension and reinforcement (not including the cost of linking to the grid); those incurred in the balancing of increased volatility in the power system; and those that may be incurred to maintain the adequacy of the power system (i.e., its ability to cover peak demand).

A recent study, which modeled balancing costs in a number of European countries[46], found that in Germany, additional balancing costs of wind power at approximately 10% penetration of electricity (i.e., more than present penetration) amounted to approximately EUR 2.5 per MWh wind).

Key Factors with Bearing on Variable RE

This section addresses factors related to the integration of variable RE (i.e., to date, wind and solar PV power) in the German power system.

1. Coordinating Planning Initiatives

Under the Energy Concept, Germany has begun a process to holistically plan the evolution of its entire energy system—a task so large that its complexity is only beginning to be understood. The process considers the transmission needs of the whole power sector, conventional and renewable, as well as other needs that will arise from a changed power portfolio, such as changes to the operation of existing power plants, and the extent to which other system resources can be used to manage variability and uncertainty.

The Energy Concept sees three trends that run contrary to the historical German model: increasing generation of electricity in coastal and offshore plants (rather than near the principal inland demand centers), increasing distributed generation from biomass and PV plants (rather than from multi-gigawatt conventional plants), and increasing interchange of electricity with the wider European area.

Before the Energy Concept, the transmission needs of wind power were assessed and managed somewhat separately from the remainder of the system. For example, a study by the German Energy Agency (DENA) in two parts (2005 and 2010) assessed the need for additional grid reinforcement and extension solely for wind power.

A narrow view will miss the benefits to the wider system of such developments, as well as the relevance of other—perhaps unexpected—energy events. For example, the policy decision to phase out nuclear power may increase imports of electricity from France,[47] altering flows throughout the German network with decreasing nuclear generation in the south and increasing wind generation in the north.

Coordination is also seen at the European level. An interconnected European grid existed before the beginning of large-scale deployment of variable RE in Europe. Collaboration among European TSOs already allows for, at least in theory, the transmission of up to three gigawatts across any network node from the Iberian Peninsula to Scandinavia, to provide reserves in case of unexpected events.

In addition, and as a result of the EU's 3rd Liberalisation Directives, the European Network of Electricity TSOs must prepare 10-year network development plans, biannually, covering the whole of Europe. The transmission needs highlighted in this case study are not legally binding on the European Network of Electricity TSOs, but they nevertheless highlight where investment is needed. Germany's importance in the heart of Europe is reflected by the fact that it is included in four of the six regional groupings.[48]

The next edition of the 10-year network development plan will be delivered in June 2012.

Accelerated, Coordinated Deployment of New Transmission

Germany has led analysis of transmission needs to manage variable renewables. Part One of the DENA wind integration study in 2005 assessed the needs resulting from government targets.[49] This was a landmark study, and it has sparked similar analysis around the globe. But, by the end of 2010, only 90 kilometers (km) of the 460 km extensions called for by the study had been completed of; and Part Two of the study, published in 2010, published additional needs. The latest DENA analysis estimates high voltage expansion needs of 3,600 km by 2020.

The German government has legislated to combat these delays. In 2009, it published the Power Grid Expansion Act (*Energieleitungsausbaugesetz*). This, which gives additional priority to extra-high voltage "electricity highway" projects to ease north/south congestion. In August 2011, the government published the Grid Expansion Acceleration Law (*Netzausbaubeschleunigungs gesetz*), which also aims to accelerate grid expansion. The law is supposed to shorten planning and permission processes from ten years to four years by bundling responsibilities at the level of the federal government and allowing for early public participation (BMU 2012).

Spurred by the European rolling transmission plan being developed by the European Network of Electricity TSOs, the German Energy Industry Act (*Energiewirtschaftsgesetz*) was recently amended to require the development of a 10-year plan for Germany. This plan will be based on annual submissions from the four TSOs, and it would serve as the basis for a legally binding Federal Requirement Plan for Transmission Networks (*Bundesbedarfsplan Übertragungsnetze/Bedarfsplang*.

Incentives for Grid Investment

TenneT faces the majority of offshore integration costs. It has stated that no more than the nine offshore wind farms currently in the development pipeline can be connected because it does not have resources do more (Platts 2011). TenneT has requested that BNetzA increase the allowable rates of return on investment to attract investors; in October 2011 it was agreed to cut the rate from the present 9.29% to 9.05% for the period 2014–2018.[50] Additionally, TenneT has proposed the foundation of a joint direct current network operator to build the direct current line for the offshore connections (TenneT 2012). Since then, grid connection of offshore wind farms to support

German offshore wind energy development is of much interest on the political agenda. In this context, BMU and Federal Ministry of Economics and Technology have set up a working group on accelerating offshore connections. Additionally, grid companies have had to wait for two years until they are able to charge for the use of the line, and so recover their costs. BNetzA is considering whether to remove what is in effect a disincentive to invest.

Other measures under consideration by BNetzA include:

- The provision of loans to cover the cost of new lines
- Permission for higher use of system charges in new extra-high voltage transmission highways, or lines featuring innovative technologies such as those described below, Section 2
- Quality criteria and mechanisms for rewarding innovative grid expansion.

In the light of delays in transmission investment, some have suggested that unbundling of transmission and generation assets under the EU Liberalisation drive may have resulted in an unexpected side effect. Large multinational financial institutions, some of whom are based offshore, now hold a significant part of transmission assets. In addition to meeting regulations, the latter will emphasize the need to maximize returns for their shareholders, which may make them less inclined to make large capital investments in the system—when needed to meet purely national government objectives—than might be the case with domestic, public or private owners.

Nodal pricing of electricity is one tool that might accelerate grid upgrades in Germany. At present, the entire country is one price area. This means that a surplus of electricity in one area while supply is tight in another is not signaled by corresponding low or high prices. Such price differences are an important signal to potential investors who, if they were to reinforce connections between two neighboring areas, would stand to profit from resulting flows.

Coordinated Planning and Connection of Variable RE Plants

Under the 2000 Renewable Energy Sources Act (EEG), developers of renewable plants on land pay only for linking to the nearest substation; the TSO is obliged to pay all deeper reinforcement costs. It then passes these on to consumers. The cost burden is different for offshore plants; grid costs outside the wind farm itself are borne by the system operator, i.e., the expensive cable link to the onshore sub-station.

The EEG commands priority and guaranteed grid connection for RE plants, which is an very important driver for deployment, particularly as most of the grid is not state-owned. Also, the EEG includes provisions that define the actual process of connection and the commitment for grid operators to enhance their grid capacity if a new plant has to be connected to the grid. However, ambiguity in the latter has led to costly delays. For example, the local distribution system operator and the developer disagree about what constitutes the most suitable grid connection point. The 2011 amendment of the EEG aims to reduce these delays by requiring the distribution system operator to provide a connection schedule within eight weeks of an application (Eclareon 2011).

Under the Energy Industry Act, a TSO is obliged to ensure that a connection for an offshore wind farm is in place by the time it is ready to generate electricity. However, delays have resulted when neither party was able or willing to start construction before the other: the developer because financing proves elusive before cable construction has commenced, and the TSO because the wind farm has not yet secured financing. To address this "chicken and egg" problem, the networks regulator, BNetzA, issued a non-binding but effective "position paper" that clarified the procedure, with which, according to a recent study by Eclareon, the majority of stakeholders appears to be largely satisfied (Eclareon 2011).

To reduce offshore connection costs, Part One of the DENA study proposed that four offshore transmission hubs take precedence in planning over potentially duplicative, individual on-shore links. This hub approach has been embraced in the Energy Concept. It is hoped that this will facilitate the process and reduce the costs of offshore connection.

Within the Energy Concept, the German government has amended the Offshore Installations Ordinance (*Seeanlagenverordnung*) to significantly simplify and accelerate the approval process for installations in the German offshore exclusive economic zone (outside territorial waters). The amendment transfers the entire permitting procedure for offshore wind power plants in these waters to a single agency: the Federal Agency for Maritime Shipping and Hydrography. This is expected to minimize inter-institutional delays.

Federal Cooperation with the States (Länder)

The 2010 Energy Concept includes a "Government-Länder Initiative on Wind Energy," which intends to improve cooperation between federal and state levels in the search for higher quality wind resources on land. This is particularly important as the majority of the best resources may already have been exploited. The initiative will include an analysis of wind energy potential and will be an important tool in the ongoing process of identifying and designating suitable new sites for onshore wind (BMU 2012). This site designation process will in turn facilitate earlier planning of additional grid resources in those areas.

2. Building Public Support for New Transmission

The fact that the German population generally favors RE, and is consequently, more willing to pay for it, does not mean that public support can be counted on for the new transmission that clean energy may imply.

The four system operators in Germany have developed three scenarios of grid expansion footprints, which are included on a special website launched by the TSOs to engage the public in the decision-making process (Lang and Mutschler 2011) before presenting the first grid development plan in June 2012.

Development of the Federal Requirement Plan for Transmission Networks will feature in-depth public consultation, and it aims to emphasize the need for transmission upgrades in the mind of the population. Possible incentives include compensating municipalities hosting new transmission corridors.

The Energy Concept 2010 announced that a publicity campaign, "Grids for environmentally sound energy supply," would be launched to build public support for new transmission projects.

New Technology to Reduce Visual Impact of Grid Upgrades and Extensions

In its 2010 study, DENA considered the use of new technologies that enable increased transfer capacity on discrete line lengths without major visual impact. Flexible line management based on dynamic monitoring of line temperatures enables the operator to get more carrying capacity out of a specific line at the very time it is needed—during high wind speeds—because high winds tend to occur with lower temperatures, at which carrying capacity increases.

Replacing existing lines with high temperature lines made of aluminum instead of copper also reduces the need for landscape disturbance, allowing up to 50% greater capacity while retaining existing pylon size. Both technologies have been demonstrated in several applications, although costs are still relatively high. Flexible alternating current (AC) transmission system devices also can enable existing lines to carry more power, while wide-area monitoring can improve their management. Finally, cabling (i.e., running transmission lines underground) is an alternative, though more costly, to unsightly new overhead lines.

All these options have been demonstrated successfully in Germany and elsewhere, although their use has yet to become widespread. For example, flexible line management has been shown to increase the carrying capacity of specific lines by up to 50% in coastal regions, up to 30% in the rest of northern Germany, and up to 15% in the south of the country.

3. Increasing Flexibility of the Power System

Storage Options under Investigation

Part 2 of the DENA study extended this outlook to 2020, with scenarios featuring new storage facilities able to absorb 50% and 100% of surplus wind power generated, as well as the value of the demand-side flexible resource. The Energy Concept includes a EUR 200 million budget for storage R&D up to 2014. Under the amended Energy Industry Act, new storage facilities are exempt from grid charges. The same is true for the EEG levy ("EEG-Umlage") (BMU 2011b). The German energy agency and partners are examining the benefits of using excess electricity production to produce hydrogen, and hence synthetic methane gas, which can be fed into (and so stored) in the gas grid. Early demonstration plants are already in use. Direct hydrogen feed-in into the gas grid may require some technical adaptation. The German gas transmission system operators estimate the costs of integrating 10% of hydrogen (by volume) into the gas transmission grid to amount to about EUR 3.7 billion, excluding potential further costs related to other system components (e.g., gas storage and distribution grids, power plants, industry) (*Netzentwicklungsplan* Gas 2012).

From 2009 to 2011, the German Space Agency, a national research center that provides a major contribution to energy research, in collaboration with several other institutes, completed the *Leitstudie*, a dynamic simulation of German electricity supply with a high share of renewables, including the importance of storage.

Flexible Power Plants

Discussions regarding adapting the regulations around combined heat and power generation to improve the flexibility of such plants are ongoing within the government. Providing an incentive to owners of such plants to include additional thermal storage is one option being examined. This would allow to some extent the production of heat and power to be "decoupled." For example, when heat demand is high but a reduction in electricity output is desirable in order to balance a variable RE surplus, the heat may still be drawn off the storage, while the actual production can be reduced. The draft law presently under discussion considers the use of grants to cover 30% of the costs of the additional heat storage (Argus 2011). The law aims to increase the role of combined heat and power plants in power production from approximately 15% in 2010 to 25% in 2020.

The Energy Concept targets the completion of fossil-powered plants now under construction by 2013, and an additional 10 GW of plant capacity by 2020. But, simultaneously the government has provided incentives in the form of a new fund to encourage this new capacity to take up the latest flexible technology, so that its output can better complement variable RE (BMU 2011b). In short, the intention is to maximize the speed at which plants can ramp up and down, resulting in minimal overlap with variable RE output (and consequently minimal unnecessary emission of CO_2) and reliable support for system adequacy.

Government and private sector partners have for many years worked on the concept of a virtual power plant—one that provides all the electricity services of a conventional power plant but consists of several different plants in different locations, each providing a part of this service. One of the most advanced developments in this area has been the combined power plant concept (*Kombikraftwerk*) developed by a consortium of variable RE producers (Kombikraftwerk 2 n.d.). This concept links wind, solar, biomass, and hydropower installations throughout the country.

4. Making Markets More Amenable to Variable RE

In addition to the FIT, one of the more important driving forces behind renewables deployment in Germany has been the priority dispatch (*EE-vorrang*) accorded to renewable power, ensuring that as much of its output as possible is used. The significance of this is considerable; since 1991 when priority dispatch was first introduced, all renewable electricity must be bought, even if this means base-load plants such as nuclear have ramp down (SOU 2007).

Until quite recently, variable renewable power plants could ignore the power market. The electricity they produced would be bought by the system operator and sold directly to suppliers. But, this changed when the Equalisation Scheme Ordinance (*AusglMechY*) of 2009 stipulated that all EEG electricity was to be sold by the TSOs on the spot markets. In 2012, the logical next step came into effect, wherein under the 2011 July EEG, RE producers could choose a "Market Premium" option over the existing FIT. By implementing the Market Premium, the government wants to stimulate learning processes with regard to electricity markets in the RE sector and aims at a more demand-responsive generation of RE.

Producers choosing to adopt this support model will sell their production, themselves, in the spot market. New companies that organize the professional selling of RE production and which are setting up huge pools of plants have been established. Dispatchable renewable electricity generators such as biomass and geothermal power plants could then have access to a potentially important revenue flow—resulting from the flexible provision of more valuable balancing power in response to system needs—alongside typical spot market prices for energy.

Protecting the Revenue of Existing Flexible Resources

As wind and solar PV electricity production increases, and because of their low marginal cost and priority dispatch, less production is needed from existing conventional plants, such as gas and coal, which have higher operating costs (mainly fuel). This is known as the "merit-order effect" (i.e., whereby conventional power plants are pushed down the order in which plants are used).

This "missing revenue" problem may adversely affect the economics of those plants to the point that owners no longer consider their continued production to be profitable and retire them from service. If this would occur, it would not only reduce the amount of flexible power on the system able to balance fluctuating variable RE output, it might also undermine the adequacy of the system (i.e., its ability to meet its peak power requirements). At present, the discussion of this very important issue is stalled because the potentially resulting market intervention is deemed profound, and it may well be an academic discussion for now. As a next step, the issue will be discussed in the context of the biannual Power Plants Forum organized by Federal Ministry of Economics and Technology. Even if fossil-fueled plants are displaced to some extent by new variable RE output, they will be needed to compensate for the nuclear power plants already retired (nearly 10 GW), alongside imports of electricity from France.

Unbundling of Generation and Transmission Asset Ownership

The on-going liberalization of the European Union's electricity markets is driving the separation of production and transmission asset ownership. This was undertaken to address the disincentive on bundled utilities to provide access to their transmission assets, which would increase competition with their generation assets.

Until recently, the "Big Four" German utilities (E.ON, EnBW, RWE, Vattenfall) also owned and operated the high-voltage transmission system. However, E.ON sold its transmission assets to TenneT, the state-owned Dutch system operator (which created TenneT TSO GmbH) in 2010; Vattenfall sold to Elia and IFC Infrastructure, an Australian Infrastructure fund, in 2010; and RWE sold 75% of Amprion in July 2011 to a financial consortium managed by Commerz Real AG, which consists of several European pension funds. In other words, system operators that do not own generation assets now control three of four of the TSO areas in Germany. In contrast, EnBW Transportnetze AG, in the southwest of the country, still belongs to EnBW, but EnBW is itself majority-owned by the state of Baden-Württemburg.

Widening the Market

For more than a decade, driven by EU liberalisation directives, European electricity markets have been moving toward greater interconnection and integration of markets. While the former has often met with very lengthy delays—new lines can take 15 years to accomplish— consolidation of the markets that depend on this grid has continued apace, particularly in recent years. While the primary objective of this consolidation was never to benefit variable RE specifically—but rather to lower costs to the consumer in an even manner—it also benefits variable RE. This is because market consolidation makes possible the trading of less-expensive variable RE output over much larger areas; it enables valuable flexible resources to flow to where they are needed most; and, potentially, it allows greater complementarity of variable RE outputs from different resources (e.g., wind power in the northern Germany with non-correlated solar PV output from Spain).

Since November 2010, Germany has been an integral part of a single electricity market that includes France and Benelux countries through a process known as "market-coupling." In essence, this means simply that cross-border transfer capacity no longer has to be acquired explicitly—through a separate arrangement from the actual acquisition or sale of energy—but is implicit in the trade itself, greatly reducing the time required to effect international trades. In effect, it encourages less-expensive (including variable RE output) electricity to flow toward higher cost areas where it will find a market; it also reduces price differences across borders.

This Central and West Europe market is linked to the Nordic market via a similar mechanism, but one in which the coupling is activated only at a certain cross-border price differential, known as the Interim Tight Volume Market Coupling. This was developed by a joint venture of German Nordic TSOs, known as the European Market Coupling Company.

Greater, Faster Trading of Power

Over the same period, spot markets (i.e., voluntary exchanges wherein electricity can be bought and sold on a range of timescales) have emerged and consolidated, providing further liquidity in the electricity market, which is beneficial to the balancing of variable RE. The European Power Exchange saw approximately 24% of German electricity trade in 2009 (Eclareon 2011), as well as trading of French, Swiss, and Austrian electricity. However, this means that in that year, three quarters of German electricity was still traded bilaterally, rather than through a pool.[51]

The European Power Exchange includes a day-ahead spot market, as well as an intraday market, on which trading volume is growing rapidly; it grew from 1.6 terawatt-hours (TWh) in 2007 to approximately 10 TWh in 2010.[52] BNetzA announced in 2011 that henceforth balancing power needed against wind forecast uncertainty would be traded on the intraday market; this will further increase volume at the expense of the—less transparent—balancing market (Borggrefe et al. 2011).

Gate closure times in day-ahead and intra-day markets have been shortened to 15–75 minutes ahead of the time the electricity is actually generated and consumed. This increases the ability of the market to optimize the level of dispatchable plant output to complement the latest forecast of variable feed-in from wind and solar PV resources.

Opening up the Balancing Market

After the spot exchanges have closed, electricity required by the system operator to balance any remaining as well as newly occurring imbalances between supply and demand can be found through the balancing market. Historically, the balancing market has been rather opaque. This was particularly a concern before system operation was unbundled from generation ownership. Transparency has increased with the introduction of a compulsory Internet platform,[53] which was introduced following a July 2005 amendment of the Electricity Industry Act and which includes details of all tenders to the balancing market, nationwide.

Between 2007 and 2010, tendering for balancing power was common to all four control or balancing areas, though the actual balancing of these areas was done separately, meaning that simultaneously balancing of energy still occurred in contrary ways (e.g., one area ramping up supply while its neighbor ramped down). This was remedied to some extent in 2010, when BNetzA stipulated a common secondary reserve market with a standardized tender (Eclareon 2011), but primary reserve (the shortest timescale) is still tendered separately in the four control areas.

In 2011, BNetzA specified new terms of tender for balancing energy to facilitate the market access of smaller energy suppliers, including consumer appliances that allow the capacity to be adjusted, electricity storage, and renewable energy systems.

- *Primary* system balancing, which must be provided in seconds to stabilize frequency deviations in the whole European network, is tendered for 1 week in advance (instead of 1 month as it was previously); the minimum tender quantity is 1 MW (instead of 5 MW); and it is now possible to pool plants.
- *Secondary* system balancing, to control differences between the different balancing areas, is also tendered for 1 week (before: 1 months), and the minimum tender quantity is 5 MW (before: 10 MW) and it is possible to pool plants.

Challenges for Neighboring Countries

Polish and Czech system operators are considering blocking action in the face of large wind-based flows into and through their systems. Poland is considering installing devices to enable this (Platts 2011). Austria, for example, buys wind power to fill its pumped-hydropower reservoirs, and 35% of electricity flowing from Germany to Austria passes through the Czech Republic.

5. Promoting Diversification of Location

Diversifying the location of variable RE plants has the effect of smoothing their aggregated output, as the solar and wind resources in different sites will fluctuate in an uncorrelated or only partially correlated manner.

The support system in Germany includes a specification with regard to the FIT, depending on the wind potential available at a specific wind turbine site. To that end, a reference turbine (including a reference site) is introduced. The energy yield of the first five years of operation of a turbine is compared to its theoretical reference energy yield using a defined calculation method. In this way, the initial (high) FIT period, which usually lasts five years, can be extended (EEG 2012). This mechanism was intended to enable wind energy development sites with lesser (but still valuable) wind resources. This may result in greater diversification of wind turbine sites, and thus, smoothing of aggregated output might result as a side effect of differentiated payments. It has probably not had practical consequences yet, though this might change with the growing willingness of the southern German states to designate site zones.[54]

Planning at state level, including wind priority sites, is more important for diversification of wind plants. In the southern states (especially in Bavaria and Baden-Württemberg), not many wind priority sites have been identified, but this is changing.[55]

6. Advanced System Operation and Grid Code Development

The task of TSOs, which manage the high-voltage grid in areas with very large shares of variable RE electricity, is increasingly complex. Very large amounts of data need to be managed and continually updated, while more dynamic management of power plants, such as re-dispatching or using curtailment, requires high-speed decision-making. For example, in June 2011, 50Hertz, the system operator in eastern Germany, inaugurated a new control center, in large part to manage the increasing complexity of managing a power system with large amounts of wind power. In planning system operation, the center uses multiple wind output forecasts in combination, running up to a 96-hour forecast horizon. These are updated twice daily (Jones 2011).

In April 2011, the Federal Network Agency (BNetzA) amended rules relating to the ability of system operators to "redispatch" power plants. To alleviate congestion in a specific line, generators on either end may be instructed by the system operator to alter their output. As a more sustainable remedy, new transmission lines are necessary in the medium term. It should be noted that another driver behind the changed rules has been the nuclear phase out.[56]

Curtailment

Serious delays to essential grid expansion work are also apparent in the increasing need to curtail wind plants in the north of the country. Though an important system management tool, the curtailment of power plants (or "feed-in management") leads essentially to the waste of what was wanted in the first place (i.e., clean energy) so it should be minimized. Indeed, its use in Germany does have a "last resort" status. Nevertheless, curtailment in 2010 increased by up to 69% over the previous year. Even if it only amounted to 0.2% - 0.4% (72–150 GWh) of total wind electricity,[57] in some northern wind farms as much as 25% of output was curtailed (Ecofys 2011). Under the EEG, curtailment losses have mostly to be compensated by the system operator, which passes these costs on to the consumer.

Until it was amended in 2011, the EEG required wind and solar PV power plants (among others) with nominal capacity greater than 100 kW to be able to be curtailed remotely by the system operator in the event of electricity surplus, and to make information about the current electricity feed-in of an installation accessible to the system operator. Failure to provide this capability would in theory result in disqualification for the FIT. However, in practice, this requirement proved difficult to apply to PV systems.

The curtailment requirement did not apply to power plants smaller than 100 kW. However, under the latest amendment of the EEG, remote control management of PV plants down to 30 kW is required (EEG 2012). It is optional for plants smaller than 30 kW, which may choose instead to have a reduced feed-in (Stetz 2011). In addition, the distribution system operator has certain other emergency powers to require curtailment by the plant owner under the Energy Industry Act.

Other Requirements of the Grid Code

From April 1 2011, all new power plants will need a certificate proving that they comply with requirements, collectively known as the "grid code."

In 2009, the German government issued a new rule (SDLWindV 2010) that requires certain other services of variable power plants, though to varying extents. These are usually referred to as ancillary services. For example, all wind power plants commissioned since 2009 must be able to remain on the grid, generating electricity to support it, in case of faults (this is known as "fault ride-through capability"). Plants must also be able to control the frequency of their output in order to protect the frequency (stability) of the network local to their connection. Plants commissioned since July 2010 are required to provide additional services.

For older wind power plants, incentives are provided to encourage them to be refurbished in order to provide some of these services. Plants commissioned before January 2009 are offered a bonus of EUR 7/MWh for five years if they completed refurbishment to provide both fault ride through and frequency control by January 1ˑ 2016 (Fichtner 2010, section 66, paragraph 8).

The speed with which solar PV power has appeared in the German electricity system—although it still has only a 2% share—has resulted in additional requirements of it. This is partly because almost all the installed PV has appeared at the low and medium-voltage grid level. Typically, the lower voltage grid network, unlike the high voltage network, is not actively managed. Since 2010, PV power plants connected to the medium-voltage grid (10 kV to 110kV) have had to comply with certain standards. These include voltage support and active power control; since 2011, they include fault ride-through capability (Ellis 2009).

Potential for Learning in Other Countries

The transmission and distribution grid in Germany evolved to serve fewer (relative to today) large, gigawatt-scale, dispatchable power plants located close to centers of high consumption. The country is evolving to be better able to manage the output and needs of a very large number of additional, geographically dispersed variable output power plants at scales kilowatts to hundreds of megawatts.

Germany has also been at the forefront of renewables deployment. It has iterated its transmission plans, connection rules, and grid codes, resulting in considerable learning and expertise. Its experiences in moving from a generator-owned, vertically integrated electricity system to one of mixed ownership and increasingly private ownership, while promoting significant use of RE and increased investment, should greatly interest most countries with power grids established by the latter half of the last century.

The BMU has been the primary driving force behind variable RE deployment and integration from the outset. Countries where the preference is to leave energy matters entirely to the market will find the German model more difficult to emulate. However, German actors as a whole also understand the importance of regulated market forces. Progress toward market coupling, though initiated by Belgium, France, and the Netherlands, has been enormously reinforced by the German presence, not least of all, because it provides a link to the Nordic market.

German experiences may be less applicable to smaller developing countries that have neither a mature grid nor the wide market opportunities provided by Germany's position at the geographical center of Europe. A fully interconnected grid may not be financially feasible in some cases.

Conclusion

With the 2010 Energy Concept, and its update in summer 2011, the German government has embraced the complexity of a fundamental shift in energy production. Germany is moving away from nuclear power toward an energy future based in large part on a fundamentally different power portfolio. The Energy Concept is intended to capture and manage this complexity that includes: fluctuating, partially predictable output; diverse location of plants, some far offshore and many others embedded in the low-voltage grid; and the resulting need for a more flexible grid that is stronger at its extremities in order to harvest all this new resource.

Germany has long been a test-bed for renewables deployment, and it has made considerable strides in integrating variable generation, though significant tasks remain. It remains to be seen whether the Energy Concept will solve the biggest challenge: rolling out and reinforcing the grid. Germany has shifted from wind-only transmission studies to the development of a legally binding Federal Requirement Plan for Transmission Networks that is hoped will resolve delays.

Other factors will be important. Public support for new transmission is a principal barrier. Local incentives, a major information campaign, and new technology to reduce the impact in the most sensitive areas are all being tried. Investment needs to be found, too. Locational/nodal pricing of electricity to reflect where congestion is highest—and therefore guide investment to where it is most needed—could help. It may also be that higher regulated allowable returns on investment may be needed to attract investment.

The priority grid connection accorded by law to RE plants has been an enormous spur to investors, as has the allocation of costs—in the case of offshore, entirely—away from the plant developer. This procedure has caused delays, but the government has acted to resolve these. Again, results remain to be seen, but accelerated offshore deployment is anticipated. Recent steps to clarify jurisdictional issues outside of coastal waters may also help. Onshore, federal cooperation with the Länder is hoped to lead to additional efficient land-based deployment.

In addition to supporting the physical ability of output to flow to the user, Germany has done much to make marketing of that energy more flexible to suit its variable nature. The feed-in premium model to be launched this year is hoped to encourage more dispatchable renewables out from under the shelter of the EEG and into the marketplace, where they can be better used to balance variability. At the same time, storage, decoupled combined heat and power plants, more flexible conventional generators, and the virtual power plant concept target further increase in flexible resources. Priority dispatch ensures that the most variable RE possible is brought to market.

Transmission and generation ownership have been largely unbundled, with benefits for competition, though new owners may also prove difficult to convince when it comes to making new investments unless their returns increase. Enabling more fluid trade north and south through market coupling is facilitating management of local concentrations of variable output. Greater liquidity on the spot markets, particularly close to the time of operation, is bringing more flexible resources to bear, revealing their value and thus boosting investment.

As complexity increases, so does the task of operating the power system. Newly independent system operators are building new control centers to plan the operation of their areas using advanced forecasting techniques. Short-term balancing, particularly to cover uncertainties in these forecasts, has become more open. Collaboration has been greatly improved among the four control areas, with more efficient use of balancing resources, although they are still managed separately.

A good measure of the flexibility of the system is the extent of curtailment of variable RE output. Curtailment is on the rise as grid congestion increases. Given the strategic energy choices of government, it should in future be minimized, even if at small energy volumes it can be considered a useful source of flexibility.

References

Argus. (2011). "German Government Proposes Changes to CHP Subsidies." Accessed April 13, 2012: http://www.argusmedia.com/pages/NewsBody. aspx?id=778733&menu=yes.

BMU (Federal Ministry for the Environment, Nature Conservation and Nuclear Safety). (2011a). "Renewable Energy Sources in Figures, National and International Development." Berlin: BMU.

BMU. (2011b). "The Energy Concept and its Accelerated Implementation." (2011). Accessed April 13, 2012: http://www.bmu.de/english /transformation_of_the_energy_system/resolutionsand measures/doc/48054.php.

BMU. (2012). "Energiewende auf gutem Weg: Zwischenbilanz und Ausblick." [Energy Concept for an Environmentally Sound, Reliable and

Affordable Energy Supply] Berlin: Bundesministerium für Umwelt, Naturschutz und Reaktorsicherheit (BMU) and Bundesministerium für Wirtschaft und Technologie (BMWi). http://www.bmu.de/files/pdfs/ allgemein/application/pdf/broschuere energiewende.pdf.

Borggrefe, F.; Neuhoff, K. (2011:). :Balancing and Intraday Market Design: Options for Wind Integration." http://www.climatepolicyinitiative.org /files/attachments/96.pdf.

Couture, T.; Gagnon, Y. (2010). "An Analysis of Feed In Tariff Remuneration Models: Implications for Renewable Energy Investment." *Energy Economics* (38:2); pp. 52-58.

[DENA] Deutsche Energie-Agentur (Germany Energy Agency) (2005). "Planning of the Grid Integration of Wind Energy in Germany Onshore and Offshore up to the Year 2010 (dena Grid study): Sumary of the Essential Results of the Study." http://www.dena.de/fileadmin/user_ upload/Publikationen/Energiedienstleistungen/Dokumente/dena-grid_ study_summary.pdf. Accessed April 17, 2012.

DENA. (2011). "Integration of Renewable Energy Sources in the Germany Power Supply System from 2015-2020 with an Outlook to 2025." http://www.dena.de/fileadmin/user_upload/Publikationen/Sonstiges/Doku mente/dena_Grid_Stud y_II_-_final_report.pdf. Accessed April 17, 2012.

Eclareon. (2011). "Integration of electricity from renewables to the electricity grid and to the electricity market. Draft Final National Report: Germany. " Berlin: Eclareon.

Ecofys 2011: Bömer, J., Burges, K., Nabe, C., *Abschätzung der Bedeutung des Einspeisemanagements nach EEG 2009 – Auswirkungen auf die Windenergieerzeugung im Jahr 2009*, January. Referenced in Eclareon, 2011.

Ellis, A. (2009). "Interconnection Standards for PV Systems." Presented to the Utility Wind Integration Group. http://www.uwig.org/pvwork/4-Ellis-InterconnectionStandards.pdf.

[EEG] Erneuerbare-Energien-Gesetz (Renewable Energy Law). (2012). Act on granting priority to renewable energy sources. http://www.erneuerbare-energien.de/files/english/pdf/application/pdf/eeg_2012_en_bf.pdf. Accessed April 17, 2012.

Eurostat. (2011). Online Eurostat database. Accessed January 30, 2011.

Fichtner. (2010). "Grid Codes for Wind Power Integration in Spain and Germany: Use of Incentive Payments to Encourage Grid-friendly Wind Power Plants" Stuttgart, Germany: FICHTNER.

IEA (International Energy Agency). (2011). *Harnessing Variable Renewables, a Guide to the Balancing Challenge.* Paris: International Energy Agency.

IEA Wind (Implementing Agreement for Co-operation in the Research, Development, and Deployment of Wind Energy Systems). (2009). "Design and Operation of Power Systems with Large Amounts of Wind Power: State-of-the-Art Report" Task 25 report, VTT Working Paper 82, Finland: VTT.

Jones, L.E. (2011). "Strategies and Decision Support Systems for Integrating Variable Energy Resources in Control Centers for Reliable Grid Operations, Global Best Practices, Examples of Excellence and Lessons Learned."

Kombikraftwerk 2 (no date). "The Combined Power Plant." Accessed April 13, 2012: http://www.kombikraftwerk.de/index.php?id=27.

Lang, M.; Mutschler, U. (2011). "TSOs Launch Joint Website for Grid Development Plan." German Energy Blog. Accessed April 13, 2012: http://www.germanenergyblog.de/?p=7922.

Platts (2011, November). "Renewable Energy Report."

Prognos 2011: EEX and TSOs solar and wind data, and load data, from 01/2011 to 12/2011. http://www.transparency.eex.com/de/daten_uebertra gungsnetzbetreiber/stromerzeugung/tatsaech liche-produktion-wind; http://www.transparency.eex.com/de/datenuebertragungsnetzbetreiber /stromerzeugung/tatsaechliche-produktion-solar; http://www.50hertz.com/de/2017.htm; http://www.amprion.net/lastverlauf-netzeinspeisung#; http://www.enbw-transportnetze.de/kennzahlen/lastverlauf/; http://www.tennettso.de/site/Transparenz/veroeffentlichungen/netzkennza hlen/jahreshoechstlast/l astverlauf-hoechstspannungsnetz .

SDLWindV (2010). Ordinance on System Services by Wind Energy Plants. http://www.erneuerbare-energien.de/files/english/pdf/application/pdf/sdl_windv_en.pdf. Accessed April 17, 2012.

SOU (Statens Offentliga Utredningar). (2007). "Grid Issues for Electricity Production Based on Renewable Energy Sources in Spain, Portugal, Germany and United Kingdom." Annex to Report of the Grid Connection Inquiry. Stockholm: SOU.

Stetz. (2011). "German Guidelines and Laws for PV Grid Integration." Presented to IEA PVPS Task 14 Meeting, October 10th, Beijing. http://apps1.eere.energy.gov/solar/newsletter/pdfs/03germanlawsand standards thomasstetz.pdf.

TenneT (2012). "TenneT Structural Solution Facilitates German Energy Transition." Accessed April 13, 2012: http://www.tennet.org/english /tennet/news/tennet-structural-solution-facilitatesgerman-energy-transition.aspx.

Weber, C. (2010), Adequate Intraday Market Design to Enable the Integration of Wind Energy into the European Power Systems." *Energy Policy* 38(2010); pp. 3155–3163.

APPENDIX E. CASE STUDY: IRELAND

Author: Jenny Heeter, National Renewable Energy Laboratory (NREL)

Introduction

Increasing penetrations of variable renewable energy (RE) concern Ireland, as it has seen increasing curtailment of wind and will see greater penetrations of RE as the country ramps up to its target of 40% RE by 2020. Curtailment of wind generation in Ireland has become more prominent in recent years; in 2009, 0.2% of total wind energy was curtailed, while in 2010, 1.2% of total wind energy was curtailed (Holttinen et al. 2011). Curtailment concerns wind generators in particular because although they are paid for what they would have generated by selling into the market, they are not compensated through Ireland's support mechanisms (Rogers et al. 2010).

In many ways, Ireland is ahead of the curve in preparing for integrating high penetrations of RE. Uniquely, the transmission operator in Ireland, EirGrid, is state-owned; for this reason, some policies discussed in this appendix may have been easier to implement. In addition, policy and market development may be less complex, given that Ireland is a small country where stakeholders are more likely to know one another.

Ireland represents an example of a deregulated market, with few imports and exports. The small system has seen a high level of variable RE penetration. Table E-1 lists key aspects of Ireland's system.

Table E-1. Key Aspects of Ireland's System

Deregulated market
State-owned transmission operator
Isolated, small system
High level of penetration

This case study examines key areas where Ireland has acted to accommodate high penetrations of variable RE. After a brief background section addressing contextual factors to Ireland's increasing penetration of RE, this appendix presents information on the following specific measures to accommodate high penetration of variable RE:

1. Coordinate and integrate planning
2. Expand access to diverse resources
3. Lead public engagement including facilitating new transmission
4. Improve system operations

Background

Variable RE, primarily wind, in Ireland has been increasing dramatically. Wind generation in Ireland increased from 0.3 billion kilowatt-hours (kWh) in 2001 to 2.7 billion kWh in 2010 (EIA 2012). At the end of September 2011, 1,585 MW of wind, 237 of MW hydropower, and 46 MW of smaller renewables[58] were operating in Ireland (EirGrid 2011). On April 5, 2010, wind generation reached 53.5% of instantaneous electricity demand on Ireland's system (EirGrid 2012).

The share of RE generation is planned to be 40% in 2020. The development of new RE is driven by EU Directive 2009/28 EC, which ensures that 16% of all energy (electricity, heat, and transport) come from renewable sources by 2020. To meet this directive, Ireland, in its National Renewable Energy Action Plan, set a target of 40% of electricity coming from renewable sources (DCENR 2011), which translates to approximately 4,600 MW of cumulative installed wind capacity in 2020 (Figure E-1).

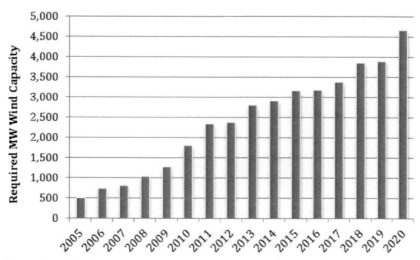

Source: Jones 2011.

Figure E-1. Cumulative installed wind capacity in Ireland to 2011 and indicative targets from 2010–2020.

Renewable generation is facilitated by the Renewable Energy Feed-in Tariff (REFIT) scheme, which was introduced in 2006 and funded in 2007. REFIT initially was designed to support up to 1,450 MW (Ireland 2010) and offered payments to onshore wind, landfill gas, biomass and small hydropower facilities (i.e., less than 5 MW of capacity). REFIT payments in 2010 ranged from €66 ($86)/MWh to €84 ($110)/MWh, depending on the technology (DCENR 2012). The first offering "REFIT 1" was closed at the end of 2009, but recently the European Commission indicated it intends to provide state aid clearance for a second REFIT round "REFIT 2". REFIT 2 will cover the same technologies as REFIT 1 and will be available to facilities that commenced construction after January 1, 2010 (DCENR 2012).

Market and Technical Background

It is important to recognize that Ireland is a small, island system that depends on energy imports (Bazilian n.d.). There is a minimal level of regional interconnection, with only three AC links (with Northern Ireland) with a capacity of 450 MW (Bazilian n.d.). Northern Ireland is connected to Scotland by a 500-MW HVDC link, and, as will be discussed later, a 500 MW HVDC link between Ireland and Great Britain should be completed by the end of 2012 (Holttinen et al. 2011).

Ireland presents a case of a government that owns the TSO. The TSO in Ireland, EirGrid, is a state-owned company established after Ireland deregulated its power sector. The EirGrid board of directors is appointed by the Minister for Communications, Energy and Natural Resources, and the company is regulated by the Commission for Energy Regulation. The System Operator of Northern Ireland (SONI), which cooperates with EirGrid, is regulated by the Northern Ireland Utility Regulator.

Five Areas of Intervention to Accommodate High RE Penetration

This section discusses how Ireland has sought to coordinate and integrate planning; expand access to diverse resources; lead public engagement including facilitating new transmission; and improve system operations. Each topic area highlights a specific approach taken by key stakeholders to address the high penetration of variable RE.

1. Coordinate and Integrate Planning

Energy ministers in Ireland and Northern Ireland have supported planning; EirGrid, in cooperation with SONI, is leading follow-on work. A number of studies have been conducted to assess impacts of high penetrations on the all-island system (Table E-1).

Energy departments in Ireland and Northern Ireland recognized the need for transmission planning, and in 2005, collaborated on a preliminary all-island 2020 vision for RE. A resulting working group recommended completing an *All-Island Grid Study*. Completed in 2008, the *All-Island Grid Study* was the first to examine future impacts of high penetrations of RE and was the result of multiple integrated work streams, including a resource assessment, network study, portfolio screening and dispatch studies, and an analysis of costs and benefits. The *All-Island Grid Study* acknowledged that there were limitations to the methodology and that future technical work was needed.[59]

Following the *All-Island Grid Study*, the *Facilitation of Renewables* studies demonstrated that significant changes would be needed to reach the renewable targets set by both governments. The Facilitation of Renewables studies were designed to identify potential technical issues related to high penetrations of renewables and develop mitigation measures to inform future operational policies, grid code and standards, and payment mechanisms for all generation (EirGrid and SONI 2009). EirGrid and SONI contracted with ECOFYS, Siemens-PTI, and Ecar to complete the work.

Table E-1. Summary of Key Studies

Study	Year Published	Primary Objective	Primary Participants, Authors
All-Island Grid Study	2008	Examine a range of generation portfolios, ability of grid to handle various penetrations of RE, investment levels required, climate and security of supply benefits	Department of Communications, Energy and Natural Resources (Ireland), Department of Enterprise, Trade and Investment (Northern Ireland), ESB International, University College Dublin—The Electricity Research Centre, Risoe National Laboratory, University of Stuttgart, RAM-løse EDB, University of Duisburg-Essen, and TNEI Services Limited
All Island TSO Facilitation of Renewables Study	2010	Identify potential technical issues and mitigation measures	
Delivering a Secure Sustainable Electricity System	Ongoing	Implement technical solutions identified in the Facilitation of Renewables Study	EirGrid and SONI

Most recently, EirGrid and SONI collectively established a work plan, *Delivering a Secure Sustainable Electricity System (DS3)* to implement the technical solutions developed in the *Facilitation of Renewables* study. The DS3 program includes 11 separate work streams: frequency control, voltage control, system services review, demand side management, grid code, performance monitoring, rate of change of frequency, model development and studies, renewable data, Wind Security Assessment Tool, and control center tools and capabilities.

DS3 has a built-in communications and stakeholder engagement plan that includes an advisory council, industry forums, and work with the existing Joint Grid Code Review Panels. Stakeholder engagement is planned throughout the DS3 process, with regular advisory council meetings, industry forums, and grid code review panels. In addition, the TSOs and regulatory authorities in Ireland and Northern Ireland are planning to meet three times in 2012.

Ireland's planning efforts have progressed from studying the potential for renewables to taking concrete steps to addressing technical issues related to the high penetration of variable RE. Energy ministers in Ireland and Northern Ireland supported the initial study, the *All-Island Grid Study*. The TSOs in Ireland and Northern Ireland built on that work, and they are implementing technical solutions.

2. Expand Access to Diverse Resources: Creation of the Single Electricity Market and Interconnection to Great Britain

Expanding markets is a key way to decrease the operational challenges of integrating variable RE into the grid. Ireland and Northern Ireland established a common Single Electricity Market (SEM) in 2007 to unite the electricity markets of the island. The expanded market creates a greater pool for balancing and enables more transmission connection to Great Britain.

SEM achieves these objectives as all electricity exceeding 10 MW sold and bought in Ireland is traded through the central electricity pool through a market clearing mechanism; no bilateral transactions are permitted outside the pool. According to the SEM Committee, SEM "promotes the interests of consumers by enabling greater competition through cost reflective prices, while also securing a diverse, viable and environmentally sustainable long term energy supply" (SEMO 2010). The SEM covers approximately 2.5 million electricity customers in Ireland and Northern Ireland (SEMO 2012).

Cooperation of regulatory agencies and transmission operators in Ireland and Northern Ireland was required to establish the SEM. The Single Electricity Market Operator (SEM-O), the organization that operates and administers the SEM, was developed through a series of memorandums of understanding. The first memorandum of understanding was signed in 2004 between the regulatory bodies in both countries: Commission for Energy Regulation (Ireland) and the Northern Ireland Authority for Utility Regulation. In 2005, a memorandum of understanding was signed by EirGrid (then ESB National Grid) and SONI, the grid operators in Ireland and Northern Ireland. When SEM launched in November 2007, EU Energy Commissioner Andris Piebalgs said, "The Single Electricity Market is important not only for the island of Ireland, but also for the European Union. This is a significant contribution to the construction of the internal energy market. A regional initiative like SEM deserves to be widely copied (EU 2007)."

The SEM is the first market in the world to operate in multiple jurisdictions and dual currencies, the Euro and Sterling. The market operation has been deemed successful, with "no major issues with regard to the operation and oversight of the market under two jurisdictions" (Conlon 2009). SEM has been compared to the Nordpool Power Exchange and the Eastern Australian markets (Conlon 2009), though it is a small market with a spot market volume of 34.6 terawatt-hours (TWh) in 2009, compared to the spot market volume in Nordpool of 285.5 TWh in the same year (Nepal and Jamasb 2011).

Addressing system curtailments was one of the most contentious issues, which was recently resolved by the SEM. A process was developed for *which* generators to curtail when the need to curtail wind generation arises. This issue was long and controversial primarily because wind generators receive an incentive payment based on the megawatt-hours they generate. The SEM considered equally distributing curtailment among all generators but ultimately ruled that the newest installed generators would be curtailed first (e.g., wind capacity installed in 2011 would be curtailed before wind capacity installed in 2010). SEM said this approach "should provide a signal to the marginal renewable plant in future years of whether it is viable to connect to the system (SEMO 2011, p. 17)." Once the 2020 RE targets have been reached, this process may be reevaluated.

East-West Interconnector to the United Kingdom

Policymakers and grid operators have recognized the need to expand Ireland's grid. Reasons for expanding go beyond accommodating higher penetrations of RE.[60] In 2006 and 2007, plans for an East-West interconnector linking Ireland to Wales developed. In 2006, the federal government requested the link be incorporated in the Commission for Energy Regulation's energy policy (Edwards 2010), and in 2007, the government's National Development Plan 2007–2013 called for an East-West connector (NDP 2007). The impact of export capability on wind curtailment levels at 38% wind penetration was modeled by EirGrid and SONI (System Operator for Northern Ireland), demonstrating that increased export capability would lower curtailment levels (EirGrid and SONI 2010).

The planning process for the 500 MW line involved 2 years of marine surveys, route selection, public engagement, route amendments, and pre-application consultation (EirGrid 2009b). The European Commission awarded the project €110 ($144) million in stimulus funds in 2010. Construction of the interconnector began in July 2010 and the interconnection is expected to be operational in Q3 2012. Unicorn Systems developed an auction management platform (AMP) to allow participants to trade across the interconnector, which went live in November 2011. During the AMP development, two workshops on access rules were held (October 2010 and November 2011), and stakeholder comments were integrated into the final rules.

3. Leading Public Engagement, including Facilitating New Transmission

Public engagement and facilitating new transmission development is particularly important to Ireland, as it is classified as an island system with little interconnection.

Public engagement is led by EirGrid and supported by the Ireland Wind Energy Association (IWEA). EirGrid has two methods of engaging stakeholders: public education efforts and outreach for specific transmission projects.

EirGrid reaches out to schools in order to educate student about the benefits of wind power. The effort focuses on educating teachers and students, not about one specific transmission project, but rather about the energy sector and infrastructure projects generally. In this way, once a specific transmission proposal is developed, the public is already aware of the issues involved. IWEA operates a separate education effort using the KidWind[61] curriculum developed by the American Wind Energy Association.

Communities in Ireland have expressed concern over both the aesthetic impacts of new transmission lines and the health impacts related to electromagnetic radiation emissions. Community opposition continues to be a challenge – as construction on the East-West Interconnector began, protesters vowed to block roads, citing opposition to the final route selected (Independent 2010a). There has also been pushback against EirGrid's paid media outreach campaign, because the company is a state-owned monopoly. In 2010, EirGrid ran 30- second advertisements on television at an estimated cost of €600,000 ($785,000) for ad development and airtime (Independent 2010b).

Facilitating New Transmission

The existing Moyle interconnector connects Northern Ireland to Britain and serves almost 4.7% of total generation capacity in the SEM (Nepal and Jamasb 2011). The link has a capacity of 450 MW (Edwards 2010).

Ireland's national grid development strategy, Grid25, provides for grid expansion over the long term (through 2025) to support the Irish government's target to increase renewable electricity to 40% of generation. Grid25 was supported by the Irish Minister for Communications, Energy and Natural Resources, Eamon Ryan but developed by EirGrid, the state-owned transmission operator. A stated goal of Grid25 was to exploit Ireland's wind and wave energy sources, as well as to facilitate transmission of RE, in line with the Irish government policy. In addition, the development of multiple lower-voltage lines could be avoided through advanced planning of a single higher-voltage line.[62] Grid25 found that there was a need for about 1,150 km (715 miles) of new circuits and 2,300 km (1,429 miles) in upgrades.

To facilitate the development of new lines, EirGrid must scope out individual transmission projects, propose them to the public and key stakeholders, and then seek approval from the Commission on Energy Regulation.

Recently, Ireland centralized the planning approval process from a local approval to a national approval. The Strategic Infrastructure Board was created under the National Planning Authority (An Bord Pleanála). The Strategic Infrastructure Board approves transmission plans on a national basis, so instead of a transmission developer needing approval from multiple local planning departments, only one approval is needed. Although some transmission developers may have had positive relationships with local planning boards that were supportive of development, the transition has been viewed favorably.

4. Improve System Operations: Advanced Forecasting

EirGrid is improving system operations by deploying a Wind Security Assessment Tool (WSAT) in its control room. The WSAT was developed with Powertech Labs to assess the maximum amount of wind generation that the system can accept, by examining voltage and transient stability analyses of transfers during normal operation (Holttinen et al. 2011). WSAT was installed and officially launched in October 2010 at the National Control Center in Dublin. There are also plans to implement the tool in Northern Ireland.

Importantly, the early involvement of system dispatchers in the development of WSAT and EirGrid's Wind Dispatch Tool resulted in faster integration of the tools into the control room (Jones 2011).

Replication of Efforts by Other Countries

Ireland can serve as a prime example of how to plan for high penetrations of RE, expand markets and balancing areas, lead public engagement to facilitate new transmission, and improve system operations. It should be noted that each country has a unique set of circumstances that may make the actions discussed here relevant or not.

Ireland is a small, island system. Similar systems should look towards Ireland's efforts to expand markets and balancing areas, if possible.

EirGrid, as a state-owned transmission operator, has been forward thinking and supportive of Ireland's aggressive RE targets. EirGrid's board is appointed by the Minister for Communications, Energy and Natural Resources.

Summary and Conclusion

Ireland has set aggressive goals for RE development and is actively working to facilitate high penetrations of variable RE. Specific measures taken in Ireland include:

1. **Coordinate and integrate planning.** Planning began with the All-Island Grid Study (2008), which was supported by energy ministers in Ireland and Northern Ireland. The Study examined a range of generation portfolios, the ability of the grid to handle various penetrations of RE, the investment levels required, as well as the climate and security of supply benefits. In 2010, EirGrid and SONI completed the Facilitation of Renewables Study, which identified potential technical issues and mitigation measures. Currently, EirGrid and SONI are implementing technical solutions identified in the Facilitation of Renewables, through the Delivering a Secure Sustainable Electricity System work program.

2. **Expand access to diverse resources: Creation of the Single Electricity Market and interconnection to the U.K.** Ireland and Northern Ireland established a common Single Electricity Market (SEM) in 2007 to unite the electricity markets of the island. The SEM was established through a series of memorandums of understanding between regulatory agencies and transmission operators in Ireland and Northern Ireland. The SEM has been working well, and recently established a new process for curtailing wind; newest installed generators will be curtailed first.

   To further expand access to diverse resources, Ireland is also constructing an interconnector to the U.K. The line will have a 500 MW capacity and is expected to go online in 2012.

3. **Lead public engagement including facilitating new transmission.** EirGrid's education outreach promotes awareness of the needs and

purposes of new transmission generally. More specifically, a national grid development strategy, Grid25, provides for grid expansion through 2025, and EirGrid must scope out individual transmission projects, proposed them to the public and key stakeholders, and then seek approval from the Commission on Energy Regulation.

4. **Improve system operations: Advanced forecasting.** EirGrid has improved system operations by deploying a Wind Security Assessment Tool (WSAT) in its control room. The involvement of system dispatchers in the WSAT development resulted in faster integration of the tool into the control room.

In the future, as Ireland's penetration of RE increases, evaluating connections with the European grid will become important. As connection becomes feasible, changes will need to be made to the SEM.

Generally, policymakers might benefit from considering how policy goals are working together. For example, in Ireland, priority dispatch legislation, market systems prioritizing the most efficient dispatch, and production-based support payments can be in conflict with each other. Through the SEM, Ireland has begun to address these issues; however, investors may not be satisfied by solutions that reduce support payments to wind generators.

References

Bazilian, M. (not dated). A Way Forward for Wind in Ireland. Unpublished draft.

Conlon, P. (2009). The Integration of Electricity Markets in Ireland under the ISO Model. http://www.esbi.ie/news/pdf/White-Paper-Integration-Electricity-Markets.pdf. Accessed February 22, 2012.

Department of Communications, Energy and Natural Resources (DCENR) (2012). REFIT. http://www.dcenr.gov.ie/Energy/Sustainable+and+Renewable+Energy+Division/REFIT.htm Accessed February 22, 2012.

Edwards, N. (2010). Interconnection with the UK and the Role of Merchants Lines. EDF Energy. http://research.edf.com/fichiers/fckeditor/Commun/Innovation/conference/16juinChatou/6NigelEdwardsE   DFEnergy.pdf Accessed February 7, 2012.

EirGrid 2012. All-Island Wind and Fuel Mix Report. Summary 2011. http://www.eirgrid.com/operations/systemperformancedata/all-island wind andfuelmixreport/ Accessed February 15, 2012.

EirGrid (2011). Inaugural Annual Renewable Report: 2010. http://www.eirgrid.com/media/Annual%20Renewable%20Report%20201 0.pdf Accessed December 14, 2011.

EirGrid (2009a). Executive Summary: Interconnection Economic Feasibility Report. http://www.eirgrid.com/media/47958_EG_Summary09.pdf. Accessed February 22, 2012.

EirGrid (2009b). East-West Interconnector News. http://www.eirgridprojects. com/media/Eirgrid%20East%20West%20Interconnector%20Newsleter%2 0De cember.pdf. Accessed February 22, 2012.

EirGrid and SONI 2010. "Executive Summary." Accessed April 17, 2012: http://www.eirgrid.com/media/Faciltiation%20of%20Renwables%20Exec utive%20Overview%20and%2 0Agenda.pdf

EirGrid and SONI 2009. TSO Facilitation of Renewables Studies. http://www.eirgrid.com/media/Facilitation%20of%20Renewables%20For um%20Session%201.pdf. Accessed February 22, 2012.

EIA (Energy Information Administration) (2012). International Energy Statistics. http://205.254.135.7/cfapps/ipdbproject/IEDIndex3.cfm? tid= 2&pid=2&aid=12. Accessed April 2, 2012.

European Union (EU). (2007). *Commissioner Andris Piebalgs welcomes the launch of the Single Electricity Market Belfast.* News release. http://europa.eu/rapid/pressReleasesAction.do?reference=IP/07/1696&for mat=HTML&aged=1&languag e=EN&guiLanguage=en Accessed February 7, 2012.

Holttinen, H., Orths, A.G., Eriksen, P.B., Hidalgo, J., Estanqueiro, A., Groome, F., Coughlan, Y., Neumann, H., Lange, B., van Hulle, F., and I. Dudurych. (2011). Currents of Change: European Experience and Perspective with High Wind Penetration Levels. IEEE Power & Energy Magazine. Volume 9. Number 6. November/December. DOI: 10.1109/MPE.2011.942351.

Independent (2010a). Protest against EirGrid power work. http://www.independent.ie/breakingnews/national-news/protest-against-eirgrid-power-work-2432598.html. Accessed February 22, 2012.

Independent (2010b). EirGrid defends €600,000 publicity campaign. http://www.independent.ie/national-news/eirgrid-defends-600000-publicity-campaign-2191859.html. Accessed February 22, 2012.

Jones, L. (2011). Strategies and Decision Support Systems for Integrating Variable Energy Resources in Control Centers for Reliable Grid Operations: Global Best Practices, Examples of Excellence and Lessons Learned.

http://www1.eere.energy.gov/wind/pdfs/doe_wind_integration_report.pdf Accessed February 22, 2012.

National Development Plan (NDP) (2007). Transforming Ireland: A better quality of life for all. http://www2.ul.ie/pdf/932500843.pdf Accessed April 4, 2012.

Nepal, R. and T. Jamasb (2011). Market Integration, Efficiency, and Interconnectors: The Irish Single Electricity Market. University of Cambridge Electricity Policy Research Group. EPRG Working Paper 1121, Cambridge Working Paper in Economics 1144. http://www.eprg.group.cam.ac.uk/category/publications/working-paper-series/. Accessed February 22, 2012.

Ravenscroft, V. (2012). PUC EIM Group: Background & Overview of Draft Market Design Effort. http://www.westgov.org/PUCeim/webinars/02-10-12/02-10-12slides.pdf. Accessed April 4, 2012.

Rogers, J., Fink, S., and K. Porter (2010). Examples of Wind Energy Curtailment Practices. National Renewable Energy Laboratory. Golden, Colorado. NREL/SR-550-48737.

Single Electricity Market Operator (SEMO). 2012. Website. http://www.semo.com/. SEMO (2010). SEM Committee Annual Report 2009. SEM-10-027. http://www.allislandproject.org/en/wholesale_overview. aspx?article=9fe266b6-27a8-4692-909e217048f9791d. Accessed February 22, 2012.

SEMO (2011). Treatment of Price Taking Generation in Tie Breaks in Dispatch in the Single Electricity Market and Associated Issues. Decision Paper. December 21, 2011. SEM-11-105. http://www.allislandproject.org/en/renewable_current_consultations.aspx?article=baec321e-5542-44d9-8fb2-491fffab7972 Accessed February 13, 2012.

APPENDIX F. CASE STUDY: SPAIN

Author: David Pérez Méndez-Castrillón, Ministry of Industry,
Energy, and Tourism

Coordinated and Integrated Planning

Policy and Planning

Spain's energy situation as well as the policies pursued in the last decades are the direct result of certain challenges: a high degree of energy dependence, a lack of sufficient interconnections (as it is almost an isolated electric system), high energy consumption per unit of gross domestic product, and high levels of greenhouse gas emissions (mostly due to a strong growth in electricity generation and to the energy demand in the transport sector). To face these challenges, energy policy in Spain (and in other European countries) has spun round three axes: security of supply, enhancement of the competitiveness of Spain's economy and a guarantee of sustainable economic, and social and environmental development.

The path followed by Spain rests on the development of policies, strategies, and instruments enabling progress all along the three challenges. In Spain, energy policy priorities have focused on market liberalization and transparency, the development of a flexible and diversified energy mix with adequate energy infrastructures, and the promotion of renewable energy (RE) and energy efficiency.

The RE energy policy proposed takes into account that Spain has one of the highest levels of energy dependence in Europe, and that the Iberian Peninsula makes up an electric system that is isolated from Europe. To this end, Spain has promoted policies enacted to support the generation and management of RE and has ended the regulatory and institutional barriers for developing RE.

The Spanish energy policy is set forth the Ministry of Industry, Energy and Tourism, with counseling by the Institute for Energy Diversification and Saving (IDAE)—a public business entity answerable to the Ministry through the Secretariat of State for Energy. The development of the RE sector in Spain is based on few but key issues: a proper policy; a mature business and industrial sector; a strong institution network; and technical support.

Implementing RE policy in Spain has been supported by:

- *Binding objectives established through energy planning* that aim to set RE targets tailored to the renewable source potential of the country
- *National renewable energy plans that were drafted to achieve these targets*
- *Economic regulatory framework and financing support schemes* and programs that allocate economic incentives through a feed-in tariff system, with a view to helping achieve technological maturity in RE; this system has been in force for the last 30 years and has been subject to ongoing improvements and modifications customized to the breakthrough in technological development; it has helped achieve high levels of penetration in clean energy, with similar unit costs.

- *Technical regulatory framework* that gives priority access to the electrical grid, and ensuring a suitable investment level in infrastructures (see "Supporting Model Grid," below)
- *Binding planning infrastructure* that has led to significant technological development, which has given way to an efficient operation of RE installations and their integration into a power system under a strict security level; the latter has involved adapting the operation of RE facilities to the safe operation of the electric network, and the running of a control center for RE (CECRE) to monitor these installations on the basis of real-time information availability; all of this accounts for the maximum integration of RE, in tune with the required level of security in the Spanish electricity system.
- A *business sector mature* enough to undertake intensive investments and to be implemented quickly
- *Institutional and technological support,* provided by a strong institutional network leading to the development of capacity building programs; the national research, development, and innovation support framework consists of research and development programs and grants; besides, the establishment of technology platforms and public research and development infrastructures has enabled to achieve a leading position in RE; this has also generated a major development in manufacturing and service companies and research centers of international prestige, such as the Center for Energy, Environment and Technology and the National Renewable Energy Center furthermore, 4-year national research and development and innovation (R&D&I) plans are programming instruments to define strategic actions spanning over priority sectors and technologies, including RE technologies.

Spain is a good example of how a supportive public energy policy has helped drive technological advancement as well as business development by creating a mature industrial RE sector and allowing the integration of substantial amounts of intermittent RE in the electricity system.

Renewable Energy Plans

Implementing a comprehensive indicative energy planning (with suitable participation, transparency, and overall periodic revisions and follow-up) is the essential starting point for any consistent energy policy.

The beginning of the development of RE in Spain began shortly after the second international oil crisis with the enactment of *Law 82/1980 on energy conservation*. The Government, through the Center for Energy, Environment and Technology, opened the Almería Solar Platform in 1981. The Almería Solar Platform was the first center for solar technology research intended to show the technical feasibility of concentrating solar energy as a power source.

These actions were followed by *Law 54/1997 on the Electricity Sector*, which deregulated the electric power market and made Spain the first European nation to introduce a legally binding renewable resource objective; 12% of the primary energy demand had to be met by 2010 using renewable sources. To this end, this law called for the preparation of a renewable energy development plan, which was approved in December 1999 as *Plan 2000-2010 for the Promotion of Renewable Energies*[63]. The plan analyzed the status and potential of these energies, fixed specific objectives for each technology and set up different measures and instruments through a system based on strategic planning (including infrastructures), sources diversification, flexibility, and monitoring.

When the 12% objective was seen as difficult to fulfill, Spain took further measures by approving the new *Renewable Energy Plan 2005-2010*[64] in 2005 as well as an action plan to improve energy efficiency in order to increase the speed at which renewable installations were to be put into service and therefore, curb the rising energy demand. The purpose of this revision was to maintain the 12% RE target by 2010, as well as add two additional targets, adopted within the European Union: 29.4% of electric power generation from renewable sources and a 5.75% increase in the use of biofuels for transport by 2010.

Drafting the RE plans has been a transparent and participatory process that embraced the autonomous communities of Spain[65], the National Commission for Energy (the national regulatory body in Spain), the Ministries of Economy, Finance and the Environment, as well as manufacturers, business associations, and the public (e.g., trade unions, non-governmental organizations).

The success of the national RE plans and integration of RE in Spain can be summarized in the following results:

- **Energy Demand Coverage**: At the end of 2011, RE covered 13.2% of final energy consumption and 33% of the total electricity production in Spain. On November 6, 2011, Spain achieved a new record when wind power provided 59.6% of electricity demand; the previous peak was 54.0%. In 2010, RE covered 11.8% of final energy consumption and 33.3% of the total electricity production in Spain (see Figures F-1 and F-2).

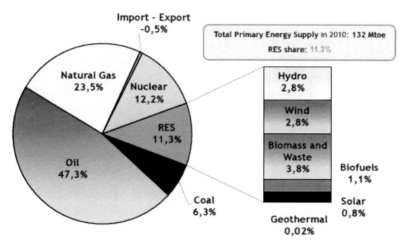

Source: Institute for Energy Diversification and Saving.

Figure F-1. RE in primary energy consumption in Spain, 2010.

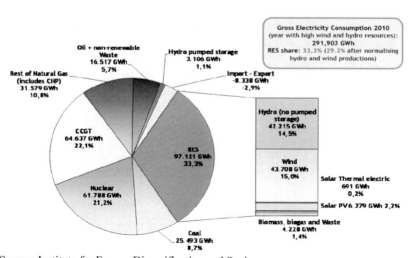

Source: Institute for Energy Diversification and Saving.

Figure F-2. RE in electricity consumption in Spain, 2010.

- **Security of Supply**: In March 2011, and for the first time, wind produced more electricity than all other technologies and met 21% of the monthly demand, setting a monthly generation record of 4,738 GWh); such generation might meet the monthly consumption of a country like Portugal (Spanish Electric System Report 2011).
- **CO_2 Emissions**: The environmental benefits associated with RE are many. From 2005 to 2010, electricity production from fossil fuels fell in Spain by 68.4% for coal and 26.35% for oil. The specific CO_2 emissions by the electric sector fell by 40% in this period (from 407 to 247 TCO_2/GWh). Spain is very close to the required level to achieve the target allocated to Spain by the European Union (223 TCO_2/GWh) (Institute for Energy Diversification and Saving).
- **Entrepreneurial Activity***: The renewable sector in Spain comprises more than 4,000 companies of differing sizes and activities, some of which are internationally renowned for their operating capacity and proprietary technology production. This approach has led to a thriving RE industry in Spain, based on technological innovation and close cooperation with the public sector.
- **Electricity Production Costs**: Costs associated with developing RE, inclusive of premiums and the cost of the different support systems, involve a leverage action onto direct and measurable benefits deriving from RE, such as the reduction of imported fossil fuels, lowering of CO_2 emissions, export of technology, and the creation of associated jobs. Additional benefits should that are more difficult to quantify include rural development. Renewable energy technologies are becoming cost-competitive and cost reductions in critical technologies such as wind and solar are set to continue.

Renewable Energy Plan 2000-2010 has been a success. Not only has it transformed Spain's energy model as planned, it has also helped develop an industry that has taken the lead in all the segments of the value chain at international level.

Network Planning

The Ministry of Industry, Energy and Tourism has published on a four-year basis its official planning documents for the coming 10 years. These grid-planning studies are based on studies of load forecast and generation adequacy. *Red Eléctrica de España* (REE) — the TSO in Spain— has developed planning tools to cope with the uncertainty of future RE generation in terms of production and location.

- **Generation Adequacy Studies in the 10-Year Horizon:** High RE penetration levels imply that generation adequacy studies become more complicated than they are with fossil fuel power plants due to the intermittent nature of RE generation. Consequently, RE production planning requires the use of multiple scenarios. In Spain, 8,760 hours are simulated per year using past RE production series, which are up-scaled to the forecasted RE installed power. Using the forecasted demand data series, power production from conventional generation can be obtained and analyzed. The impact of high RE levels in the production required from conventional generation implies that thermal power plants must be able to cope with the variability of RE production. When this is not feasible, the TSO must rely on imports from, and exports to, neighboring systems. However, when the level of interconnection is not enough, as it is the case of Spain, RE curtailment will be the only solution. A main objective in the planning studies of the TSO in Spain is to propose mechanisms to minimize those RE curtailments. New pumping stations, new interconnections, and new fast response power plants (i.e., those using open-cycle gas-turbine or OCGT technology) can be considered and evaluated.
- **Grid Adequacy Studies in the 10-Year Horizon:** Grid studies with high RE penetration are not conceptually different from those with only conventional power plants in the generation mix. Nevertheless, the distributed nature of RE in a country like Spain implies that the location of these plants is not as well-known as that of conventional power plants, which introduces even more uncertainties into the grid planning process. In Spain, the total RE installed capacity by technology is known, but its extent by region is not. To address this uncertainty, the TSO in Spain analyses the maximum capacity that can be installed on each bus and each significant electrical region of the grid, taking into account the potential simultaneity of RE production in different areas, identifying all significant potential congestion problems and selecting the best cost-benefit solutions.

Proper parallel grid planning is almost as important as setting objectives; in this sense, network planning in Spain aims to minimize the cutback on RE.

Enabling Larger Balancing Areas or Markets to Help Manage Variable Generation

Market integration is essential to achieving the goals of having a sustainable and secure electricity system with a great potential for renewables integration. The Spanish electricity market was merged with the Portuguese market in 2007, forming the Iberian electricity market (MIBEL). The internal bottleneck between Spain and Portugal is handled by splitting the market into Spanish and Portuguese price areas. Bilateral trading is only possible before the day-ahead market closes. With MIBEL, any consumer in the Iberian Peninsula can acquire electrical energy under a free competition regime, from any producer or retailer that acts in Portugal or Spain.

MIBEL has established a physical support platform for the Iberian regional market, supported by the development of transport infrastructures and the articulation of energy planning and transport networks. It also joins the legal and regulatory frameworks of the economic conditions to take part in MIBEL, as well as the system operational procedures. As interconnection capacities to neighboring countries are quite low, balancing has to occur mainly within the market area.

Moving gate closure nearer the corresponding operating period results in more accurate forecasts and better-performing schedules in systems with significant penetration of intermittent renewable sources. Various existing mechanisms help achieve shorter gate-closure terms in accordance with the large benefits available. Also, Spain has developed intraday markets, which help shorten the gate-closure period. This market design has facilitated the market integration of RE.

MIBEL largely contributes to the integration of larger amounts of intermittent renewable energies. It is a good example that may serve other regions in the world as a real and tested approach to market integration.

Developing and Integrating Advanced Forecasting Techniques into Grid Operations

An important characteristic of the Spanish power system is it has dispatch priority in scheduling available power plants to meet system load requirements, even as RE production presents *high levels of intermittency and uncertainty*. In this context, the availability of flexible conventional power plants such as open cycle gas turbines or flexible combined cycle units has enabled the power system to respond to RE variability.

However, the integration of RE production has also been conditioned by the particular characteristics of the electricity system in Spain. Isolation is one of its most relevant structural features. From an electric point of view, *Spain has one of the lowest interconnection ratios in the European Union.*[66] This lack of sufficient interconnection capacity has prevented the Spanish system from taking advantage of cross-border exchanges for the integration of RE, as cross-border exchanges enable electricity exports when the surplus of renewable production cannot be properly dispatched in the system, thus diminishing RE curtailments and increasing the overall efficiency. This means special attention must be paid to coordinating, aggregating, and controlling the overall production that is fed into the grid because a certain volume of non-RE units must also be dispatched to fulfill with security and technical constraints. That RE plants tend to be far more distributed and dispersed than conventional power plants complicates this task.

In response to this challenge, the system operator in Spain established a control center of special regime, the Spanish Control Centre of Renewable Energies (CECRE), whose objective is to monitor and control RE production, maximizing its production while ensuring the safety of electrical system. CECRE was established in June 2006 as wind generation started to become a relevant technology in the Spanish electrical system. It is composed of an operational desk where an operator continuously supervises RE production.

Renewable energy control centers collect real-time information and channel to the CECRE. To minimize the number of points of contact dealing with the TSO, the renewable energy control center acts as the only real-time speaker with the TSO. The control center also manages the limitations established by set-points, and they are responsible for assuring than the non-manageable plants comply with them.

The use of information and communication technologies has been a key factor to achieve the success of the control center. The great potential of information and communication technologies to monitor and control energy and to dematerialize physical equipment, represents both an immediate and long-term solution to coping with increasing world energy consumption and managing variable availability of renewable sources (in order to achieve an sustainable economic model).

These control and supervision schemes lead to improved security and effectiveness in system operation; they allow the substitution of permanent or long-lasting production hypothesis and preventive criteria for real time production control, and they thus allow higher energy production for the same installed capacity and a more efficient real-time operation of power plants.

Management Tool: The main tool used by CECRE's operator to carry out these tasks is named GEMAS, the Spanish acronym for the *maximum admissible wind generation* that the power system can accommodate. GEMAS accesses the real-time information received in CECRE and uses it to determine whether the present generation scenario is admissible for the system due to each of the following criteria:

- The fault ride-through capabilities of generation plants connected to the network through power electronics will not cause an inadmissible simultaneous disconnection of generation.
- Congestions that cannot be solved by reducing other types of generation do not appear, in either the base case or the N-1 case.
- System balance can be achieved while maintaining an appropriate level of downward reserves.

GEMAS was designed taking into account that the operator must be able to create, manage, and activate a plant rapidly as situations may arise in which returning the system to a balanced N-1 secure state as soon as possible might be necessary. Because more than 800 wind parks are installed in the Spanish peninsular system, they must be as managed as automatically as possible. The reliability of the tool is a crucial issue as the failure to deliver limitations to the RE control centers could result in a significant decrease of the security of supply.

Energy Forecasting tools, SIPREÓLICO and SIPRESOLAR: The SIPREÓLICO is REE's internal wind prediction tool. It has been in continuous development since 2002, and its purpose is to generate hourly wind forecasts until hour 24 of day D+1 for each wind park connected to the system. These forecasts are also complied to obtain a wind forecast in each network node and region as well as the total wind forecast. Additionally the SIPREÓLICO generates a probabilistic forecast that is used to calculate the required level of running reserves and a 10-day forecast of the total wind energy production in the system. Its results are updated hourly. A complimentary tool to SIPREÓLICO allows modeling the wind power forecasts into future network scenarios in order to analyze and coordinate maintenance activities or to forecast the future power flows in the system, which depend greatly on the wind production. SIPRESOLAR is a solar forecasting tool that delivers hourly energy forecasts for the next 48 hours in every solar thermoelectric and solar photovoltaic plant installed in the system. It was recently put into service.

CECRE is the first national control center (Figure F-3) in the world devoted to monitoring and controlling RE, and it has been a key factor in maximizing RE integration while assuring the overall security of the electricity system. This worldwide[67] pioneering renewable control center has made it possible for more than 50% of electricity demand to be met in several hours only with wind energy contribution—an especially impressive feature in a country like Spain, with an isolated electricity system.

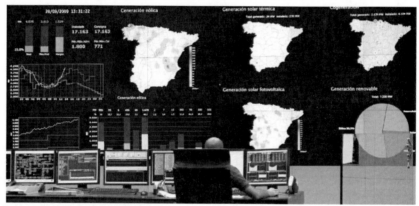

Source: Red Eléctrica Española.

Figure F-3. Main Control office in CECRE.

Supporting Model Grids

Revising codes to address concerns about variable generation (e.g., concern about frequency control and other disruptions to network stability) allows both hardware and procurement agreements to be designed in advance to support the power system, and reduces the financial burdens of retroactive requirements. Supporting a model grid code can serve as a guide for each system to evaluate changes that are needed. For these reasons, Spain has created several procedures to address some of the technical challenges in terms of grid security by accommodating RE.

- **Regulation of Access and Connection to the Network**: RE technologies open the production market to new players, who must compete with incumbent energy producers; as a result, objective access and connection network codes are needed to ensure the development of RE. In this area, REE, the Spanish TSO—an independent body from the companies operating in the electric sector (ownership unbundling model) that is responsible for granting access and connection permits to the high-voltage network, ranging between 230 kV and 400 kV—is noteworthy. Prohibition of access is strictly restricted to causes related to the system's operation security; in case of conflict, the regulator is in charge of solving it.

- **Technical Operation**: As participation of RE technologies in the generation mix increases, technical requirements in the electric system operation must be strengthened. In Spain, requirements with respect to reactive power control and responses to voltage dips in the network were first required of conventional generation facilities, then more recently of the wind, and finally of the largest photovoltaic installations. To prevent possible loss in cascade of wind generation, a procedure for responding to voltage dips was approved in 2006; in 2010, this obligation was extended to photovoltaic installations of more than 2 MW; related to a range of reactive power control, a range of power factor is established for RE with penalties in case of no compliance.

- **Operation Security of the Electricity System**: RE power plants with installed capacities greater than 1 MW must provide real-time telemetry each 12 seconds to REE, and voltage control (or power factor control) following the orders of the TSO. The regulation also requires non-manageable RE generation to be connected to a generation control center and receive set-points from the TSO.

Red Eléctrica de España, which was created in the 1980s, has been important in providing access and connection to the network and guaranteeing third-party access to the grid under equal conditions.

Conclusion

After the successes of 2000-2010 in Spain, RE is no longer a minority element in the system but rather is one of its basic components. Spain is ready to comply with European commitments and go beyond its frontiers to cooperate in the deployment of clean energies.

Having achieved a mature, qualified sector that can compete in domestic and international markets, energy policy in Spain is ready to ensure the most effective and optimal integration of renewable generation into the electricity system, and it is helping reduce emissions and increase energy independence.

To achieve a mature, qualified, and competitive RE sector, economic systems must be established to promote the cost reduction and competitiveness of RE technologies while boosting distributed generation and consumption. For example, the Spanish Government is developing a regulation for deploying a net-metering[68] system that will allow restraining energy demand on the system; this represent a re-focusing both generation and demand management rather than just generation.

Having supported deployment of the RE sector for nearly 25 years, Spain now has a mature, competitive RE sector. In fact, Spain is the leading thermal electric producer in the world; the third largest mini-hydroelectric power producer in Europe; the second largest installed-capacity wind-energy power in Europe and the fourth worldwide; and the second largest installed-capacity photovoltaic power both in Europe and in the world.

In short, Spain's success with RE penetration has featured:

- Early promotion of RE, which helped win the race toward mature technology
- Massive integration of renewable electricity into the grid under strictly controlled conditions, in an almost completely isolated country
- An essential role played by skillfully using information and communication technologies and the latest technology, such as CECRE, which the Spanish system operator created
- Economic regulation through a feed-in tariff system, acting as a lever for investments
- Smart technical regulation that prioritizes RE access to the electricity grid
- Adequate investment in research and development activities
- Long-term transmission network planning and comprehensive coordination between the various administrations
- Development of a mature industrial sector
- Close institutional cooperation and development of capacity building programs

References

[IDAE] Institute for Energy Diversification and Saving. (2011). "Employment Associated with the Promotion of Renewable Energies." Madrid: IDAE. Last accessed April 16, 2012: http://www.idae.es/index.php/mod. documentos/mem.descarga?file=/documentos_11227_e5_em   pleo   A 08df7cbc.pdf.

Spanish Electric System Report. (2011). Accessed April 16, 2012: http://www.ree.es/ sistema electrico/pdf/infosis/Avance REE 2011.pdf.

APPENDIX G. CASE STUDY:
UNITED STATES—WESTERN REGION
(COLORADO AND TEXAS)

Author: Jenny Heeter, National Renewable Energy Laboratory (NREL)

Introduction

Variable renewable energy generation strains grid operations; however, mechanisms exist to accommodate high penetrations of renewable energy. To date, regional experience in the United States has highlighted best practices in accommodating high penetrations of variable renewable energy (RE), but further action will be needed as RE penetrations increase nationwide.

Renewable energy generation in the United States is expected to increase by approximately 110 million megawatt-hours (MWh) through 2020. The U.S. Energy Information Administration (EIA) forecasts that 250 million MWh of non-hydropower RE will be generated by the electric power sector in 2020 in the United States, compared to the nearly 140 million MWh that were generated in 2010 (EIA 2012).[69] However, the level of renewable penetration varies greatly regionally. This case study presents experience from Colorado, where renewable energy accounted for 7.0% of total generation in 2010, and Texas, where wind generation alone represented 7.8% of total generation in 2010. (DOE 2011)

This case study highlights experience in two U.S. markets: the western United States (with emphasis on Colorado), and Texas. These jurisdictions provide examples of both deregulated markets (Texas) and traditionally regulated markets (Colorado). Examining these markets sheds light on four key actions described in the synthesis report.

Following a background section describing the renewable and regulatory landscapes in the western United States and Texas, this appendix presents information on the following specific measures to accommodate high penetration of variable RE:

1. Coordinate and integrate planning
2. Lead public engagement, including facilitating new transmission
3. Develop rules for market evolution that enable system flexibility
4. Expand access to diverse resources
5. Improve system operations

Table G-1 describes market structure, grid operation, and current renewable penetration in Texas and Colorado.

Table G-1. Key Aspects of Texas and Colorado Related to Renewable Integration

Texas	Colorado
Deregulated market	Regulated market
Few imports/exports	Few imports/exports
Small system	Part of large grid, but little balancing occurs
High level of penetration	Medium level of penetration

Background

Texas and Colorado have seen high levels of RE penetration, and they project additional RE to come online in future years. In Texas, data on wind generation are available from the Electric Reliability Council of Texas (ERCOT), which covers 85% of the state's load. Wind generation in ERCOT has grown from approximately 2.3 million MWh annually in 2002 to 24.7 million MWh in 2010. On a capacity basis, wind energy in ERCOT represented less than 2% of total energy through the mid 2000s. In 2010, monthly generation from wind reached more than 12% of total energy in ERCOT in April and November (Figure G-1).

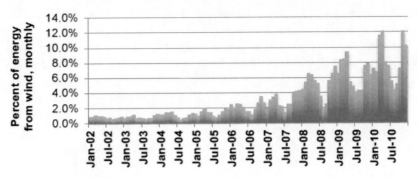

Source data from ERCOT (2011).

Figure G-1. Percentage of energy from wind in ERCOT, monthly, 2002–2010.

In Colorado, Xcel Energy, the state's major investor owned utility, announced that on October 6, 2011, 55.6% of the electricity consumed in its service territory in Colorado came from wind. (Jaffe 2011)

This case study highlights how increasing penetrations of RE have been accommodated under two different electric market structures. In Texas, markets are deregulated, while in the western United States and Colorado, markets are regulated at the state and federal levels. Deregulation in Texas began in 1995, when the Public Utility Commission of Texas required non-discriminatory access to the transmission network. Because ERCOT's grid does not cross state boundaries, it is not subject to regulation by the Federal Energy Regulatory Commission (FERC). Colorado, on the other hand, has a traditionally regulated market. Investor-owned utilities (IOUs) in the state are subject to oversight by the Colorado Public Utilities Commission. IOUs in the state are fully integrated, supplying generation, transmission, and distribution to customers. Transmission activities that cross Colorado state boundaries are subject to FERC regulation.

Five Areas of Intervention to Accommodate High RE Penetration

This section discusses how regions in the United States have sought to coordinate and integrate planning, lead public engagement including facilitating new transmission, develop rules for market evolution that enable system flexibility, expand access to diverse resources, and improve system operations. Each topic area highlights a specific approach taken by key stakeholders to address the high penetration of variable RE.

1. Coordinate and Integrate Planning: Creation of the Competitive Renewable Energy Zone (CREZ) Process

Competitive renewable energy zones (CREZ) in Texas highlight the importance of aligning timescales of generation and transmission. Texas' advanced planning efforts helped foster development of wind in the state. Originally, wind developers identified the need for new transmission as an issue after heavy development in the McCamey area. The McCamey area had excellent wind resource, but it was served by weak transmission. In 2002, 758 MW of wind were interconnected to a substation with only 400 MW of transmission (Sioshansi and Hurlbut 2010). Wind developers began developing in other parts of the state and went to the state legislature to seek a solution.

Table G-2 compares expected future renewable energy development in Texas and Colorado.

Table G-2. Future Renewable Energy Development in Texas and Colorado

Texas	Colorado
Renewable portfolio standard obligation	Renewable portfolio standard obligation
2015 target of 5,880 MW already achieved	2020 target of 30% retail sales (IOUs) 2020
Goal of 10,000 MW	Estimated to be ~2,700 MW in 202
Additional low-cost wind likely to come online	

To get transmission built, stakeholders identified two primary solutions: 1) delegate authority to a centralized institution to implement transmission planning and 2) align planning timescales for generation and transmission. A key component to aligning planning timescales for generation and transmission was the need for transmission developers to receive cost recovery assurance.

The approach surfaces in multiple avenues. The Public Utilities Commission of Texas (PUCT) identified the need for a CREZ approach but lacked the legal authority to implement the solution. They obtained this new legal authority through legislation. Key stakeholders included the PUCT, the state legislature, large industrial customers, wind developers, lobbyists and non-wind generators. Wind developers and their lobbyists advocated for a new process, while large industrial customers and non-wind generators resisted modifications because of cost. Stakeholder dialogue was important to identifying the need for a new process.

Ultimately, in 2005, the Texas legislature passed Senate Bill 20, establishing Texas's renewable energy program and directing the PUCT to develop CREZs. The PUCT designated zones in 2008, and a number of transmission projects were selected to transmit wind power from the CREZs to the eastern, more populated area of the state. The transmission projects are expected to transmit approximately 18,500 MW of wind (PUCT 2011).

A key factor in the success of the CREZ process is the alignment of planning timescales for generation and transmission. Senate Bill 20 guaranteed that transmission lines did not need to receive a "used and useful designation" before being built and that transmission costs would be absorbed by all ratepayers (spread over the whole load). As a result, transmission development could begin before wind generation, allowing transmission to be available for future wind development.

No other approaches were considered; however, as the PUCT implemented the CREZ approach, there were discussions about the level of wind generation for which the CREZ process should plan. It was ultimately determined that the CREZ process would support approximately 18,500 MW of wind development. CREZ transmission projects are expected to be complete by year-end 2013.

The success of the CREZ process can be evaluated by considering the extent to which generation and transmission-planning timescales have been aligned. To date, more transmission capacity has been built than is being used by wind generation; however, the lack of wind development is due to many factors, including decreases in natural gas prices and an expected expiration of the federal production tax credit.

Because the CREZ process was a new, unique development in Texas, lessons have been learned. Other countries or regions seeking to replicate this model should seek to build transmission "justin-time" for wind development; if transmission is developed too quickly, ratepayers might lose support specifically for underutilized lines or generally for transmission development. A second challenge in the CREZ process is the planning horizon. Although transmission planning requires long planning horizons, as they become longer, more assumptions need to be made. Finally, ERCOT has seen that it is now spending resources to solve technical issues that industry traditionally solved. For example, ERCOT is analyzing sub-synchronous interactions from series-compensated transmission lines on power electronics-based devices, such as wind turbines, and is investigating the implications of operating large amounts of power electronics devices in areas with weak grids (Lasher 2012).

2. Lead Public Engagement, including Facilitating New Transmission

The CREZ process exemplifies a quick method for gathering public input into transmission siting. The law implementing CREZ in Texas specified that the PUCT must issue a final decision on an application for new transmission line serving a CREZ in six months. The transmission developer must propose alternative routes to the PUCT and host public meetings; the PUCT ultimately decides which route to approve.

In a state such as Texas where land is primarily owned by private individuals, public input into route siting is important. The PUCT specifies that at least one public meeting must be held if 25 or more people live within 300 feet of the centerline of a transmission project 230 or fewer kV, or within 500 feet of the centerline of a project greater than 230 kV. For many CREZ line proposals, multiple public meetings are held.

The format for the public meeting is also important. Utilities in Texas found that input was gathered more effectively using an open house format in which several stations staffed with experts were set up around a room than with a general question and answer session. In Texas, the open house format gave private property owners an opportunity to learn about proposed routes, engineering designs, environmental concerns, and other issues, and allowed them to provide feedback on proposals through a multipage questionnaire and by interacting with staff. The open house format gave private property owners an opportunity to identify sensitive areas (e.g., private landing strips and cemeteries) not in the public records and suggest modifications. This interaction created a learning opportunity for transmission developers.

With 3,000 miles of new CREZ transmission lines planned, certainly not every landowner would be satisfied with the final route selection, but everyone was given an opportunity to provide input. Landowners were able to become formal parties to transmission cases at the PUCT ("intervene") or to file comments in either support of or opposition to a proposed line ("protest"). In one contentious case, more than 10,000 landowners intervened in a proposal for a new transmission line of fewer than 200 miles that ran through scenic hill country. Landowners in the region, many of whom were wealthy and politically connected, objected to the degradation of their land. Open houses for this line drew more than 3,400 people, and ultimately the PUCT approved a route following an interstate highway and required the use of monopoles near cities along the route (Doan 2011).

The public outreach process in CREZ was expedited in part because the need for the lines was not allowed as an issue of intervention. Therefore, the public component was focused on *where* the lines were to be built. Through legislative mandate, the line approval was accelerated to six months, but public open house style meetings provided the PUCT with evidence from public landowners to make a decision on route siting.

3. Develop Rules for Market Evolution that Enable System Flexibility: Use of Demand Resources in Texas

Allowing demand resources to provide ancillary services can increase the flexibility of the grid. When the Texas market was restructured in 2002, stakeholders were looking for a way to continue to allow demand customers to participate in providing services. Previously, demand customers were operating on interruptible tariffs, but those tariffs were canceled due to restructuring of the market. Due to the large amount of industrial load in Texas, demand resources are plentiful, because industrial loads have large demand that may be flexible.

ERCOT procures 2,300 MW of responsive reserves, though load participation is capped at 1,150 MW. ERCOT is considered a leader in allowing demand response to participate in the ancillary services market, and it has been cited as being "as successful as any market in the integration of interruptible loads into markets for ancillary services" (Zarnikau 2010, p. 1537).

In a February 2008 event, the successful use of demand response—then termed Load Acting as a Resource (LaaR)—was apparent when a net load event[70] occurred. Within 10 minutes of the event, demand response supplied 1,108 MW; within 12 minutes, it supplied 1,200 MW (Figure G-2 and Ela and Kirby 2008). According to ERCOT (2008), "It appears to be the deployment of LaaRs which halted frequency decline and restored ERCOT to stable operation."

The under-frequency relay response provides near-instantaneous response, such that when a large drop in capacity occurs, the grid recovers in less than one second in many cases (Wattles 2011). Since 2006, demand response has been deployed 21 times (Wattles 2011).

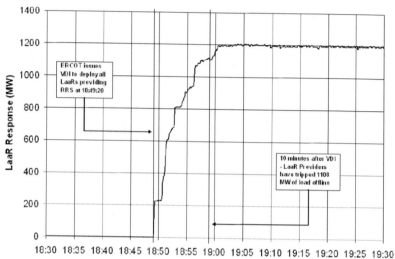

Source: ERCOT 2008.

Figure G-2. ERCOT load acting as a resource responsive reserve deployment, February 26, 2008.

4. Expand Access to Diverse Resources: Proposal for an Energy Imbalance Market in the Western United States

The design of electricity markets impacts the ability of systems to integrate high penetrations of variable RE. An energy imbalance market (EIM), currently proposed in the western United States, would re-dispatch energy on a regional, least-cost basis, with "energy imbalance" defined as the difference between actual and scheduled production and use. An EIM provides a number of benefits, including 1) reduced need for balancing reserves, resulting in operational cost savings and 2) decreased risk of one particular balancing authority depleting balancing reserves (Mariner 2011). Creating a larger geographic footprint would spread the variability of the wind across a larger area because variability declines when spread over a larger system. In addition, the creation of faster markets using EIMs could lower integration costs; In the United States, ISOs and RTOs that typically operate in sub-hourly markets see lower costs than do many nonISOs or non-RTOs, which schedule on an hourly basis (DeCesaro and Porter 2009).[71]

The Western Energy Coordinating Council (WECC) helped develop a high-level design specification and evaluate the EIM proposal,[72] but WECC is neutral as to the development decision or the organizational identity of the

EIM market operator.[73] Ultimately, the regulated utilities will need approval from state PUC regulators to address cost recovery for the EIM development.

Utilities have been involved with the development of the EIM; however, perspectives vary depending on the regulatory exposure of the utility, the level of RE penetration, and the current cost of energy. Generally, small public utilities served entirely or partially through a wholesale allocation from the federal power marketing administrations have resisted supporting an EIM more than large investor-owned utilities. This is possibly because many large investor-owned utilities have renewable portfolio obligations and are directly paying the cost of inefficient markets.

The development of an EIM has required stakeholder dialogue, primarily to educate stakeholders about the effort. In addition, education has provided information to relieve fears about market development and FERC jurisdiction. Some utilities were concerned that expanding the wholesale market would subject them to additional regulation. In the United States, FERC has jurisdiction over interstate transmission and wholesale energy transactions in interstate commerce. FERC has tried to give reassurance that an EIM would not alter the fundamental regulated (or non-regulated) nature of wholesale market participants, but some small non-jurisdictional utilities remain uneasy.

Cost represents another concern about developing an EIM. A study by WECC attempted to calculate the costs of developing an EIM, but the calculated range was too large to allow for a meaningful comparison relative to benefits (Ravenscroft 2012). Southwest Power Pool and CAISO (California ISO) are developing new cost estimates at the request of state public utility commissions in the Western Interconnection, based on a strawman market design that stakeholders developed. An analysis of the benefits to individual balancing areas is also being developed by the National Renewable Energy Laboratory (NREL).

Another approach that could be considered is a full locational marginal pricing (LMP) market. Southwest Power Pool has an EIM and is moving toward a full LMP market, but that was not seriously considered for the western United States. Given the operating culture and jurisdictional concerns in the western United States, getting people on board with an EIM should be easier than it is with a full-blown RTO.

If an EIM is adopted in the western United States, its success will need to be evaluated. A new group, the Public Utility Commissions EIM Group, which has representatives from 12 western states, hosted three webinars in February 2012, in preparation for an in-person meeting in May 2012. Not all PUC members are supportive of an EIM, though a few state commissioners have expressed interest. Although such results are anticipated, whether an EIM in the western United States would lower costs for RE integration remains to be seen.

An EIM could be implemented in other countries. Much like multiple states negotiating on a voluntary basis, multiple countries or regions could develop an EIM to link jurisdictions. An EIM may be most appropriate when trying to link regions without creating a full LMP market. When taking this action, other countries or jurisdictions would want to ensure there is support from regulators early in the process.

5. Improve System Operations: Development and Integration of Advanced Forecasting Techniques into Grid Operations

Developing advanced forecasting techniques and integrating them into grid operations can reduce the amount of system flexibility needed to integrate variable RE generation. In Texas, the specific need to integrate advanced forecasting into grid operations was highlighted during the net load event in February 2008.

It is important to note that forecasting ability in Texas was delegated to ERCOT after the PUCT identified that gaming was occurring by generators; some wind generators realized they could receive payments for backing down generation when transmission systems were congested (Zarnikau 2011). However, forecasting services were not integrated into ERCOT's operations immediately. In February 2008, ERCOT called for an Emergency Electric Curtailment Plan due to an imbalance of generation and load. One component in the imbalance was the assessment of wind power availability; other contributors included the unexpected loss of conventional generation and the quicker-than-expected evening load ramp up (Ela and Kirby 2008). At the time, ERCOT was procuring forecast services that very accurately predicted wind power; however, the wind forecast had not yet been integrated into ERCOT's system (Ela and Kirby 2008). The 2008 event prompted faster integration of the contracted forecasting service.

ERCOT improved forecasting methods by upgrading to a more advanced wind-forecasting model. In 2010, the ERCOT Large Ramp Alert System was implemented. ERCOT and AWS Truewind, a third party forecast developer,[74] designed the tool to help prepare for large and sudden changes in wind production, like those that occurred in the February 2008 event. Qualified scheduling entities are required to use the ERCOT day-ahead wind forecast for the first 48 hours of their current operating plans; consequently, the forecast is used in both day-ahead and hour-ahead reliability unit commitment studies.

Qualified scheduling entities can provide lower values to adjust for any unreported availability, but ERCOT monitors this activity to ensure the best forecast is used (D'Annunzio 2012).

In Colorado, a slightly different approach was taken. The state's largest utility, Xcel Energy, began working with the National Center for Atmospheric Research (NCAR), a federally funded research and development center, in 2008 to improve forecasting. The resulting forecasting system provides rolling 15-minute interval forecasts 3 hours into the future and hourly interval forecasts for the remainder of a 168-hour interval. The forecasting tool produces forecasts that are 35% better than those produced by former methods. In 2010 alone, the new forecasting system was estimated to save Xcel Energy $6 million (Snider 2011). The technology transfer company for NCAR, Global Weather Corporation (GWC), has contracted with Xcel Energy to provide forecasting services, and GWC markets the system to other interested parties. Key stakeholders in Colorado were a utility (Xcel Energy) and the federally funded research and development center (NCAR).

Benefits of developing and integrating advanced forecasting can be seen in both Texas and Colorado. Both states considered the reliability benefits of advanced forecasting as well as the resulting cost savings from improved forecasting. To realize the benefits of advanced forecasting, other countries should ensure that new systems are fully integrated into grid operations; if grid operators cannot access advanced forecasts, they will be of little use to grid operations.

Replication of Efforts by Other Countries

Texas has been compared to both the United Kingdom and Australia due to its isolated nature. Because the grid in Texas is primarily isolated, it is not regulated at the federal level unlike other areas in the United States. Actions in Texas could be mimicked by other nations.

The western market in the United States is composed of many states, and regional market development is often similar to countries having to negotiate with each other. Because there is no central RTO in the western United States, any market changes must be made on a consensus basis. The role of an energy minister in such cases would be to develop diplomatic measures and aid in negotiating consensual frameworks. Regional collaboration can be essential, as some policies may not be of importance to all but would benefit regions overall.

Summary and Conclusion

Regional experience in the United States has highlighted best practices in accommodating high penetrations of variable RE. Specific measures taken in Texas, Colorado, and the western United States include:

1. **Coordinate and integrate planning: Creation of the competitive renewable energy zone (CREZ) process.** After heavy wind development occurred in an area served by weak transmission, stakeholders developed the CREZ process to centralize transmission planning and align planning timescales for generation and transmission. It is important to note that ratepayers absorb all transmission costs without needing a "used and useful designation" by the PUCT. The process has been successful in developing transmission; however, other countries developing transmission "just-in-time" for wind development should note that if transmission is developed too quickly, ratepayers might lose support for the underutilized lines specifically or transmission development generally.

2. **Lead public engagement, including facilitating new transmission.** The CREZ process established an accelerated time schedule for approving new transmission lines. Texas has requirements for public meetings, and transmission developers found that open houses were more effective at gaining feedback into siting proposals than were traditional question and answer sessions. In addition to participating in open houses, landowners can formally intervene in cases at the PUCT or provide comments in support of, or opposition to, proposed transmission lines.

3. **Develop rules for market evolution that enable system flexibility: Use of demand resources in Texas.** After transitioning to a restructured market in 2002, stakeholders sought ways to continue to allow demand customers to participate in providing grid services. Because of the large amount of industrial load in Texas, demand resources are plentiful, as industrial loads have large demand that may be flexible. A Load Acting as a Resource (LaaRs) program was developed and successfully used during the February 2008 net load event, when demand resources were able to supply 1,200 MW within 12 minutes.

4. **Expand access to diverse resources: Proposal for an energy imbalance market in the western United States.** An energy imbalance market (EIM), proposed in the western United States, would re-dispatch energy on a regional, least-cost basis, with "energy imbalance" defined as the difference between actual and scheduled production and use. Extensive stakeholder dialogue to educate stakeholders is occurring. WECC has facilitated initial discussions on developing the EIM, but the market development proposal might occur outside WECC and will include approval from state PUC commissioners. The Public Utility Commissions EIM Group is preparing stakeholders for a meeting in May 2012, by developing a strawman market design and coordinating studies of market costs and benefits.

5. **Improve system operations: Development and integration of advanced forecasting techniques into grid operations.** In Texas, the specific need to integrate advanced forecasting into grid operations was highlighted during a net load event in February 2008. New forecasting services have been contracted but not integrated into ERCOT's operations. The 2008 event prompted quicker integration of the contracted forecasting service. In 2010, ERCOT implemented an additional tool, the ERCOT Large Ramp Alert System, developed by AWS Truewind and ERCOT.

    In Colorado, Xcel Energy began working with NCAR in 2008 to improve forecasting. The new forecasting system provides 15-minute interval forecasts 120 hours into the future and produces forecasts that are 35% better. In 2010 alone, the new forecasting system was estimated to save Xcel $6 million (Snider 2011).

References

D'Annunzio, C. (March 7, 2012). E-mail communication. Electric Reliability Council of Texas. Austin, Texas.

DeCesaro, J. and K. Porter. 2009. Wind Energy and Power System Operations: A Review of Wind Integration Studies to Date. NREL/SR-550-47256. December.

Department of Energy (DOE). (2011). 2010 Renewable Energy Data Book. http://www.nrel.gov/analysis/pdfs/51680.pdf. Accessed December 16, 2011.

Doan, L. (2011). Texas approves route for largest line assigned to LCRA as part of $5B build-out. SNL. http://www.snl.com/interactivex/article.aspx? id=12233226&KPLT=6 (Subscription required). Accessed February 27, 2012.

Ela, E. and B. Kirby. (2008). ERCOT Event on February 28, 2008: Lessons Learned. National Renewable Energy Laboratory. NREL/TP-500-43373. July. http://www.nrel.gov/wind/systemsintegration/pdfs/2008/ela ercot event.pdf. Accessed December 16, 2011.

Energy Information Administration (EIA) (2011). Annual Energy Outlook 2012 Early Release, Renewable Energy Generating Capacity and Generation. http://www.eia.gov/oiaf/aeo/tablebrowser /#release=EARLY 2012&subject=10-EARLY2012&table=16-EARLY2012&region=0-0& cases=full2011-d020911a,early2012- d121011b Accessed February 22, 2012.

ERCOT (2008). ERCOT operations report on the EECP event of February 26, 2008. http://www.ercot.com/meetings/ros/keydocs/2008/0313/07_ERCOT _OPERATIONS_REPORT _EECP022608_public.doc Accessed February 2, 2012.

ERCOT (2011). Generation by Fuel Type 2002–2010. http://www.ercot.com /content/news/presentations/2011/GenerationByFuelType 2002-2010.xls

GE Energy/NREL (2010). Western Wind and Solar Integration Study. http://www.nrel.gov/wind/systemsintegration/pdfs/2010/wwsis final report.pdf

Jaffe, M. (2011). Xcel sets world record for wind power generation. The Denver Post. November 15, 2011. http://www.denverpost.com/breaking news/ci_19342896. Accessed December 16, 2011.

Lasher, W. (February 20, 2012). Email. Electric Reliability Council of Texas. Austin, TX.

Mariner Consulting Services, Inc. (2011). Why an Energy Imbalance Market Will Make the Western Interconnection More Reliable. http://www.westerngrid.net/wpcontent/uploads/2011/07/Why-an-EIM-Will-Make-the-WI-More-Reliable-072811.docx Accessed February 22, 2012.

Porter, K. and J. Rogers. (2010). Status of Centralized Wind Power Forecasting in North America. NREL Subcontract Report NREL/SR-550-47853. http://www.nrel.gov/docs/fy10osti/47853.pdf. Accessed December 15, 2011.

Public Utilities Commission of Texas (PUCT). (2011). "CREZ Progress Report No. 5 (October Update)." http://www.texascrezprojects.com/page 29603432.aspx Accessed December 16, 2011.

Sioshansi, R. and D. Hurlbut. (2010). "Market protocols in ERCOT and their effect on wind generation." Energy Policy, 38(7), pp. 3192–3197. doi:10.1016/j.enpol.2009.07.046

Snider, L. (2011). Boulder's NCAR helps Xcel Energy better forecast wind generation. Daily Camera. November 10, 2011. http://www.dailycamera.com/boulder-county-news/ci_19312301

Wattles, P. (December 8, 2011). Personal communication and email. Electric Reliability Council of Texas. Austin, TX.

Zarnikau, J. (2010). Demand participation in the restructured Electric Reliability Council of Texas market. Energy 35 (2010) 1536–1543. doi:10.1016/j.energy.2009.03.018

Zarnikau, J. (2011). Successful renewable energy development in a competitive electricity market: A Texas case study. Energy Policy 39 (2011) 3906–3913. doi:10.1016/j.enpol.2010.11.043

APPENDIX H. SUMMARY OF STAKEHOLDER FEEDBACK

This appendix summarizes feedback from five stakeholder events to present results of this analysis.

World Future Energy Summit (WFES). Abu Dhabi (January 17, 2012)

1. Approximately 20 participants—mainly from technical institutes, governments, and international agencies—attended.
2. The value of this analysis of policy lessons and messages for ministers and their staff on high RE penetration was strongly endorsed.
3. Recommendations for enhancements to the analysis included:
- Address some of the key higher-level questions that governments have about high RE penetration (e.g., at what level can the grid easily accommodate RE, what is the cost of integrating intermittent resources)
- Develop concise and simple messages on these questions and the opportunity for increased RE use

- Cover all RE technologies, not just wind; especially need to explain the issues associated with higher levels of solar integration
- Dissect the role of feed-in tariffs versus feed in premiums
- Conduct more analysis of the role of storage (e.g., when and where in the system it is most needed, how to integrate it, what are the trends)
- Identify effective approaches for public education about the value and opportunity for increased RE use and the potential for scale up and how to overcome public resistance
- Include lessons from smaller–scale, isolated or mini grids and implications for smaller countries; highlight key lessons from very high level of RE penetration in some islands and elsewhere
- Describe interactions with natural gas and approaches to allow plans to incorporate gas and RE
- Share best available information on the costs of intermittency
- Examine policy roles at three levels of integration: grid, markets, and systems
- Evaluate and describe interplay with energy efficiency and demand response
- Make available the case studies for review and feedback
- Consider developing improved tools to simulate impacts of high RE penetration and optimize system design (could be an open source competition)

Paris (January 25, 2012) and Davos (January 28, 2012)

All stakeholders were very complimentary of the work, and they believed it is valuable and important to both deliver key messages to the energy ministers at the third Clean Energy Ministerial (CEM) and to continue the work. Stakeholders strongly agreed the ministers should take a critical leadership role in engaging in a robust public process. Other suggested key messages included the following: some of the draft eight categories are overlapping, and oversimplification may be misleading but the interrelationships among key action areas are important; and the case studies offer considerable value. Some stakeholders recommended narrowing to the number of key actions from five to three such as

1. Commission comprehensive assessment of current technical, institutional, human capital and market status

2. Conduct visionary studies of high or higher RE penetration, including all the elements identified.
3. Invest in capacity building and institutional formation and learning through international partnerships for planning, analysis, market operations, regulation, among others

Additional suggestions included a systems solution, crosscutting initiative at the CEM-level and perhaps a larger international partnership on integrating renewable energy, including linkages with IEA, REN21, International Renewable Energy Agency, and others. Participants at the Davos meeting from China, India, South Africa, and the United Kingdom were uniformly complimentary of the work, and they emphasized the importance of ministers leading a robust public process and they uniformly supported an ongoing high-level international effort to share best practices. They suggested a strong linkage to Sustainable Energy for All[75] was possible.

New Delhi (CEM-3 Prep Meeting) February 1, 2012

Approximately 30 participants from CEM countries, international organizations, the private sector, and technical institutes participated in a roundtable dialogue on the analysis of policy best practices and lessons with measures for achieving high RE penetration in electricity systems. This included representatives from Australia, Brazil, Denmark, Finland, Germany, India, Spain, United Arab Emirates, United Kingdom, United States, IEA, International Renewable Energy Agency, NREL, and Brookhaven National Laboratory. The results of case studies of policies for high RE integration for Australia, Denmark, Germany, Ireland, Spain, and the United States were presented along with initial recommendations for actions that governments could take to further deepen RE penetration.

Comments and Recommendations from the Participants

Framing of Recommended Actions and Case Studies
- Present recommendations for action by governments, the private sector, and system operators
- Describe how actions contribute to issues of greatest concern to Ministers (e.g., energy security, economic development). It will be

especially important to identify how these actions can help enhance energy security.

- Additional recommended actions should be provided at higher levels of penetration with this effort seeking to identify how leading countries can be even more ambitious in deepening renewable energy deployment

- Need to clarify what is meant by the low, medium, and high levels of penetration in the summary table

- Highlight the value and need for comprehensive and integrated approaches (e.g., goals and action plans, intergovernmental coordination, public-private collaboration, monitoring and evaluation). Such comprehensive approaches have been instrumental to success by the leading countries. The report should also present case studies of these integrated approaches for countries such as Spain, Denmark, and Germany.

Specific Comments on Proposed Actions

- Highlight also the need for transparency on electricity pricing for Independent Power Producers

- Along with establishment of goals and action plans, the report should also note that governments also need to clearly define who has authority for each of the key actions.

- Describe the role for priming markets with subsidies that are then phased out in a transparent and timely manner

- Note the need for policies to be designed to be flexible to changes in prices and other market conditions

- Highlight the role of governments in partnering with the private sector to advance development and demonstrate of key technologies and tools (e.g., storage, forecasting)

Recommendations and Interests in on Ongoing Collaboration

- The CEM could help facilitate professional exchanges between countries (e.g., study tours), including exchanges for system operators and for policy officials. Denmark, Spain, and Germany already host many visitors and would be glad to further expand such peer learning. Australia also offered to share their experiences through such exchanges.

- This effort could also establish networks of different types of actors (e.g., system operators, forecasters, grid code officials, private sector developers and suppliers) to share experiences and foster peer learning and also to inform policy officials of what types of actions are needed by governments.

- On-going collaboration will be most effective if ministers at third CEM meeting provide a clear mandate and communicate this to their key stakeholders
- The government of Brazil would be glad to share experiences in Brazil with scaling up of wind generation and also would like to learn from others on how to manage higher levels of wind integration
- India is particularly interested in learning about effective approaches for development of transmission infrastructure and for integration of high levels of wind generation in the grid
- It is important to further define how this effort can be closely integrated with other CEM initiatives and related international projects on this topic and to develop a well structured collaborative framework that builds on these on-going activities. This includes for example further dialogue across Clean Energy Solution Center, International Smart Grid Action Network, and the Multilateral Solar and Wind Working Group on a collaborative approach.
- This effort currently has a very broad scope. Any collaborative work proposed to continue after the third CEM meeting should have a well-defined scope and priority set of specific activities (e.g., initial activities on priority topics of common interest, which could then be expanded as appropriate in future years).

Structure of Dialogue at the third CEM Meeting

- Need to consider whether the dialogue should request a specific mandate from ministers and if the recommendation will be an on-going collaborative project across current initiatives or a new initiative on this topic. Several at this meeting expressed a preference for a collaborative project rather than another new initiative.
- In addition to a summary of the overall results, it would be helpful to have a couple leading countries provide very brief remarks (e.g., 2–3 minutes) each highlighting the most important success factors in their cases with high RE penetration.
- India expressed interest in sharing their experiences at the third CEM meeting with rapid ramp up of solar generation.

Comments on Other Issues
- The report should further describe key barriers to high RE penetration.
- The figure given for 30% of electricity generated from RE in Spain is incorrect and the source used for this figure is not reliable.

Washington, DC March 5, 2012

NREL convened a stakeholder forum on March 5 with the Department of Energy, International Energy Agency, Center for Strategic and International Studies, and World Economic Forum in support of the Clean Energy Ministerial High Penetration of Variable Renewable Energy analysis. The forum highlighted key lessons from the case study analysis on effective policies, regulatory programs, and market designs that countries can use to increase the penetration of variable RE into the power sector. Participants were supportive of the analysis and recommendations and voiced support for ongoing CEM led activities in the area of system level solutions.

End Notes

[1] Variable renewable energy is defined as renewable energy that is not stored prior to electricity generation: it includes primarily wind and solar PV energy technologies but also technologies such as tidal power and run-of-river hydropower.

[2] The Clean Energy Ministerial, launched in 2010, is a high-level forum to promote policies and programs that advance clean energy technology, to share lessons learned and best practices, and to encourage the transition to a global clean energy economy. For more information, see http://www.cleanenergyministerial.org/.

[3] Such technologies include active voltage control and regulation, primary frequency regulation, controlled ramping, interaction with system operators, and fault ride-through capability.

[4] For more information on RETI, see www.energy.ca.gov/reti.

[5] Unpublished work by David Olsen, Western Grid Group

[6] Tools such as Structured Public Involvement offer a process for quantifying community values (Jewell et al. 2009).

[7] Limits are set on line capacity to avoid overheating; lower ambient temperatures and wind can cool the lines, allowing higher limits during these periods.

[8] WECC has been tasked with determining the cost allocation of broadening interconnections, despite that WECC was institutionally designed to ensure that NERC reliability standards are implemented.

[9] For more information, see www.renewables-grid.eu

[10] For more details on WECC and ERCOT, see pages 71-72 in Osborn, D. et al. (2011). "Driving Forces Behind the Wind." *IEEE Power & Energy*. (9:6).

[11] See, for example, FERC Order No. 890 on interregional planning requirements.

[12] Such as final reports of NERC's integration task forces

[13] See, for example, NERC Integration of Variable Generation Task Force 1.4, Flexibility Requirements and Metrics for Variable Generation: Implications for System Planning Studies.

[14] Net load refers to electricity demand minus electricity supplied by variable RE and hence the electricity that must be supplied by other resources.

[15] NERC's (2011) Integration of Variable Generation Task Force 2.4 also recommends consideration of negative pricing.

[16] For example, Europe's creation of the ENTSO-E, which is creating a ten-year network development plan. The ENTSO-E has significantly increased common network planning among European TSOs.

[17] EIA aggregates solar, tide and wave generation into a single category; however, it is likely that solar constitutes the majority of generation in this category

[18] Three packages of work were published by AEMO. Lessons learned from international wind integration studies: http://www.aemo.com.au/planning/0400-0051.pdf, wind integration in electricity grids market simulation studies: http://www.aemo.com.au/planning/0400-0056.pdf and wind integration in electricity grids historical wind data analysis: http://www.aemo.com.au/planning/0400-0057.pdf.

[19] In Australia, solar and wind power are known as intermittent because of the intermittent nature of their "fuel" resource.

[20] The final Regulatory Investment Test determination and guidelines can be found at http://www.aer.gov.au/content/index.phtml/itemId/730920.

[21] The ESIPC was established after the privatization of the South Australian electricity industry. All South Australian participants, including the network service businesses, were sold to private companies. The ESIPC provided the South Australian government with a formal "observer" with the responsibility for liaising between industry, the market operator and the government. It particularly provided an oversight on transmission planning to ensure that issues important to the development in the state were raised, discussed and resolved appropriately.

[22] A working group known as the Wind Energy Technical Reference Group provided an industry reference point for the Australian Wind Energy Forecasting System. Terms of reference for the reference group can be found at http://www.aemo.com.au/electricityops/0260-0005.pdf.

[23] A presentation describing the Australian Wind Energy Forecasting System can be obtained at http://www.aemo.com.au/electricityops/0260-0007.pdf.

[24] For more information, see http://www.aemo.com.au/electricityops/0269-0001.pdf.

[25] Solar PV currently shows very strong growth due to a very favorable support scheme and decreasing PV module prices. Hence, solar PV might become a significant electricity source during the summer period in Denmark in year 2020.

[26] The drop in 2006 reflects the fact that year saw below-average wind speeds; capacity installed remained unchanged.

[27] Greennet Study led by the Energy Economics Group of the Technical University of Vienna and summarized in Holttinen, et al. (2009)

[28] Sweden had also been involved in the initiative but withdrew in 2010.

[29] According to Wind Barriers (2010), the connection lead time in Denmark averages 2.1 months, compared to an average of 25.8 months in the European Union.

[30] Roughly ten times the consumption of the average domestic consumer

[31] Denmark consists of two bidding areas: East and West Denmark.

[32] Estonia is included.

[33] German participants, from April 1, 2011, have been able to trade on Elbas up to 30 minutes ahead of delivery to give them greater opportunity to remedy imbalance in their trading positions.

[34] ENTSO-E 2010

[35] There is also a reserve-capacity market, though payments are presently extremely low, given large supply and reduced demand for that service.

[36] Negative spot prices were introduced in the Elbas Intraday market in April 2011.

[37] The report Strategies and Decision Support Systems for Integrating Variable Renewable Energy Resources in Control Centers (Jones 2011) provides an excellent overview of system management practices in Denmark.

[38] Demand, minus variable RE output, in megawatts

[39] Only those wind farms that have been developed under a call-for-tender

[40] Up regulation is when reserves are needed to provide additional energy to balance the system.

[41] KfW Bankengruppe, Allianz SE, Deutsche Bank AG, and DZ BANK AG

[42] http://www.erneuerbare-energien.de/files/pdfs/allgemein/application/pdf/eeg_2012_bf.pdf.

[43] 15 or 30 years for hydropower plants

[44] KfW (Reconstruction Loan Corporation) is owned 80% federally and 20% by the länder (state).

[45] "Evaluation of KfW programmes for renewable energies: important contribution to the energy turnaround," November 21, 2011. Kfw press release. http://www.kfw.de/kfw/en/KfW_Group/Press/Latest_News/PressArchiv/ 2011/20111121 54587.jsp

[46] Greennet Study led by the Energy Economics Group of the Technical University of Vienna, summarized in Holttinen, et al. (2009).

[47] Although in February 2012 the opposite occurred; Germany exported more electricity to France. (http://www.reuters.com/article/2012/02/14/europe-power-supply-idUSL5E8DD8 7020120214)

[48] The four areas are: North Sea; Baltic Sea; Continental Central South; and Continental Central East. The remaining two are Continental South West, and Continental South East

[49] These amounted to 274 km to 2007, plus 552 km to 2010, plus 416 km to 2015, including both extensions and upgrades.

[50] For instance, the Bundesverband der Energie und Wasserwirtschaft (BDEW) energy lobby cites the 10% level in other EU countries.

[51] Including that paid for under the Renewable Energy Law, and traded directly between the system operators and suppliers.

[52] According to Weber (2010), there exists an intra-day market trading potential of some 17 TWh. The remaining 7 TWh potential is unused because of transaction costs, market dominance, and trade-offs with the balancing market.

[53] http://www.regelleistung.net/

[54] Email communication with Prognos analysts, February 2012.

[55] Ibid

[56] It is also around such large centralized power plants that European grids have developed; they are not optimized for large volumes of distributed generation.

[57] BNetzA 2010 reported 0.1%.

[58] Small renewables are defined as small hydropower, land-fill gas, and renewable forms of combined heat and power.

[59] For example, the studies used steady-state calculations, and for this reason, dynamic issues such as frequency stability and transient stability were not considered.

[60] Other benefits cited include "enhanced security of supply, increased competitiveness, [and] reduced production costs" (EirGrid 2009a)

[61] The KidWind curriculum can be found here: http://learn.kidwind.org/.

[62] The report explains that one 400 kV line can carry three times the capacity of a 220 kV line.

[63] Royal Decree 436/2004; Royal Decree 2351/2004; Royal Decree 661/2007; Royal Decree Act 6/2009; Royal Decree Act 14/2010; Royal Decree 1614/2010.

[64] Plan de Energías Renovables 2005- 2010. http://www.idae.es/index.php/mod.documentos /mem.descarga?file=/documentos_10359_Plan_de_Energias_Renova bles_2005_2010_9da32b5e.pdf

[65] Spain is organized by a decentralized administrative system divided into 17 Autonomous Communities, each of which has jurisdiction in regionally and locally policies.

[66] The Iberian Peninsula has a very low electricity interconnection capacity compared with the rest of Europe. The existing interconnections between Spain and Portugal under the MIBEL framework do not facilitate the integration of intermittent generation produced in Spain (as Portugal is not interconnected to any other country). For this reason, interconnections between Spain and the rest of Europe through France are essential.

[67] International Energy Agency (IEA). *"Energy Policies of IEA Countries – Spain. 2009"*. http://www.iea.org/textbase/nppdf/free/2009/spain2009.pdf.

[68] Net-metering consists of an electricity balance system for renewable electric power to facilitate distributed generation for self-consumption from RE sources. The main objective of net-metering is that end users only pay for the energy they really consume

[69] Based on EIA's Annual Energy Outlook 2012 Early Release reference case, which generally assumes current laws and regulations but does not take into account potential future laws and regulations.

[70] A "net load" event occurs when a gap grows between available supply and current demand.

[71] Benefits of balancing area cooperation are further addressed in the Western Wind and Solar Integration Study.

[72] In addition, the Department of Energy provided some funding for technical work.

[73] WECC's purpose is to ensure federal reliability standards, not to address market and cost recovery issues, which complicates the development of an EIM.

[74] See Porter and Rogers (2010). AWS Truewind provides numerical weather model data and statistical prediction procedures to ERCOT.

[75] For more information, see http://www.sustainableenergyforall.org/.

In: Integrating Variable Renewable Energy ...   ISBN: 978-1-62808-572-3
Editors: S.B. Taylor and P.R. Young   © 2013 Nova Science Publishers, Inc.

Chapter 2

U.S. RENEWABLE FUTURES IN THE GCAM*

*Steven J. Smith, Andrew Mizrahi, Joseph Karas
and Mayda Nathan*

ABSTRACT

This project examines renewable energy deployment in the United States using a version of the Global Change Assessment Model (GCAM) with a detailed representation of renewables, the GCAM-RE. Electricity generation was modeled in four generation segments and 12-subregions. This level of regional and sector detail allows a more explicit representation of renewable energy generation. Wind, solar thermal power, and central solar PV plants are implemented in explicit resource classes with new intermittency parameterizations appropriate for each technology. A scenario analysis examines a range of assumptions for technology characteristics, climate policy, and long distance transmission. We find that renewable generation levels grow over the century in all scenarios. As expected, renewable generation increases with lower renewable technology costs, more stringent climate policy, and if alternative low-carbon technologies are not available. The availability of long distance transmission lowers policy costs and changes the renewable generation mix.

---

* This report, PNNL-20823, was released by Pacific Northwest National Laboratory, October 2011.

1 INTRODUCTION

This report documents the modeling approach and an overview of results from a detailed study of renewable energy in the United States using a research version of the Global Change Assessment Model (GCAM-RE). GCAM-RE contains a detailed representation of renewable energy supply for the United States embedded within the GCAM, a global, integrated assessment model of energy, agriculture, land use, and climate change.

1.1. Overview of GCAM

The analysis for this paper was conducted using the GCAM-RE, which is based on the GCAM 2.0 integrated assessment model (Brenkert et al., 2003; Kim et al., 2006). GCAM is a dynamic-recursive, partial-equilibrium model that links representations of global energy, agriculture, land-use, and climate systems. GCAM runs in 15-year timesteps from 1990 through 2095, and has 14 regions, one of which is the United States. While the present study focuses on dynamics within electricity supply sectors, the model calculates equilibria in each time period in all regional and global markets for energy goods and services, agricultural goods, land, and, where applicable, greenhouse gas (GHG) emissions. GCAM-RE contains detailed representations of wind, solar PV, solar CSP, and solar hot water.

Exogenous model inputs include service demand drivers (population and economic growth), exhaustible and renewable resource supplies, and characteristics of technologies involved in the production, transformation, delivery, and final consumption of energy. Multiple technologies may compete to provide any energy good or service, and market share is allocated to competing technologies on the basis of relative costs, using a logit choice mechanism (Clarke and Edmonds, 1993). The cost of each technology is calculated as the sum of levelized capital and O&M costs and fuel or resource costs. Capital and O&M costs are exogenous, and fuel costs are calculated from exogenous efficiencies and endogenous fuel prices. In the case of geothermal energy, the costs are calculated based on a combination of regional, exogenous supply curves that represent "resource" costs and technology. Model assumptions for scenario variables and technologies not investigated in this study can be found in Clarke et al. (2008a).

Figure 1. U.S. sub-regions used to break-out regional electricity generation.

1.2. Model Configuration

For the analysis conducted in this project, the model was configured with 12 sub-regions, as illustrated in Figure 1. Sub-regions generally followed NERC region boundaries, with some larger NERC regions sub-divided to better capture regional heterogeneity. Within each subregion, four electricity generation segments were implemented (Wise and Smith 2007: base load, intermediate, sub-peak, and peak (Figure 2). We find that four segments are sufficient to capture the overall dynamics of electricity supply and demand over an average year. The segments are defined by the number of generation hours they contain in a load-duration curve. Fossil-fuel technologies are assigned to these segments based on average capacity factors. Base load plants have the highest capacity factors, with generation in this segment favoring plants with low fuel costs, with higher capital costs tolerated since this can be amortized over a larger number of operating hours. Peaking plants, in contrast, might operate a few hours a day for a few weeks a year. Operation under these conditions favors low capital costs, even if fuel costs are higher due to lower efficiencies.

A general translation between the operational categories used here and seasonal/time of day operation is as follows: base load technologies represent plants that operate in all seasons and around the clock. Most coal-fired and all nuclear power plants are base load plants. Intermediate load plants operate during the daytime through early evening. In the U.S., natural gas plants are often used for this segment. Sub-peak plants might operate only a few hours a day in most seasons, if at all, and all day in summer-peaking regions. Peaking plants in warm regions will operate on hot summer days, during afternoon through early evening. Natural gas and oil-fired turbines have traditionally been used for this purpose.

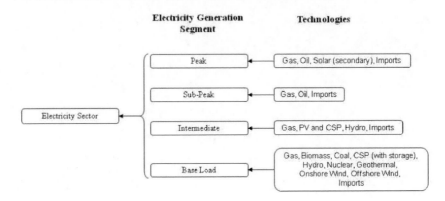

Figure 2. Electricity generation segments.

1.3. Renewable Generation

Fossil-fuel and nuclear power plants can, in principle, be located where convenient (although often adjacent to a source of water for cooling) with fuel transported, usually over long distances, to the plant. Most renewable technologies have the opposite character; generation must be located at the location of a suitable resource, and the resulting product, such as electricity, must then be transmitted to load centers. Further, for many renewable technologies, generation can vary by time of day and season, an attribute often described as intermittency. Representation of renewable resources within energy-economic models must take these attributes into account.

Renewable generation within each sub-region was calibrated in 1990 and 2005 to EIA state-level generation data, with generation pro-rated by population for states that cross NREC regions. Because this project was focused on renewable generation, and in order to simplify model set-up, generation for non-renewable resources was not calibrated at the sub-region level. Instead, non-renewable generation in each region used the average U.S. generation distribution as a starting point.

Each renewable technology supplies electricity to one or more electric load segments. Renewable generation does not always fit neatly into categories defined in terms of capacity factor. The GCAM currently treats geothermal and wind as base load technologies (as wind plants operate in all seasons; see below). CSP thermal and PV plants are treated as intermediate technologies, with an option for some contribution to peak and sub-peak. Additional demand can be met if electricity storage is paired with PV. Future

CSP plants with large amounts of thermal storage can contribute to base load generation (Zhang and Smith 2007). Rooftop PV will compete with grid-produced electricity for supplying electricity to buildings, with its higher levelized costs offset to some degree by the fact that it incurs no transmission and distribution costs or energy losses.

In the default model configuration used in this study, we account for regional differences in renewable resources by representing renewable generation in 12 U.S. sub-regions. Renewable generation within each sub-region is assumed to be available to serve loads within that subregion. While we also incorporate the additional transmission cost of connecting wind or CSP solar plants to the grid, we do not model any additional transmission investments that might be needed to augment transmission grids within a sub-region to accommodate additional renewable generation. We consider inter-regional renewable transmission as an explicit grid scenario, as described below.

1.3.1. Onshore Wind

Wind generation is represented as a set of wind resources and wind technologies, following Kyle et al. (2007). Wind resources are characterized in terms of area, distance to grid, and wind speed. The wind resource data are derived from the 50-meter wind resource map developed by the U.S. Department of Energy's (DOE) Wind Program and the National Renewable Energy Laboratory (NREL)[1]. Wind data are processed by wind class to produce a resource curve in terms of available area as a function of distance to the transmission grid.

Not all windy land can be used for power generation. Wind resources are excluded based on land-use, land-cover, population density, and an overall landscape scale constraint. Wind resources are processed to exclude protected areas, waterways (including buffer areas), and half of Department of Defense (DoD) and national forest lands. Forested areas are excluded except for an allowance for 25% of forested land with class 5 and above winds (used to represent ridge top lands). An exclusion based on population density was also applied to simulate limits on turbine placement due to visual and noise impacts. A progressively increasing exclusion starts at 5 people per km^2, with no turbine placement allowed at population densities above 325 people per km^2. Finally, we assume that no more than 50% of the land in any half-degree grid cell can be used for wind turbines, which is a simple approach to the "suitability factor" used in some wind resource estimates (Hoogwijk et al. 2004). This can be considered an intermediate assumption for land-suitability.

The impact of land suitability assumptions is examined in more detail in other work in progress (Zhou et al., *in preparation*).

Wind generation is represented by a set of technology objects, one for each wind class (class 3 – 7). Wind turbines are represented as idealized Rayleigh-Betz (R-B) turbines, characterized by a turbine rating and blade diameter. The output from the ideal turbine is reduced by three factors: a de-rating that represents real world turbine performance, conversion and turbine interaction losses, and an availability fraction. Wind speed at hub height is estimated from the wind resource data using a power law relationship between wind speed and height.

Intermittency

Wind generation varies by time of day and location, although when averaged over large areas this variation is reduced. There are a number of consequences of this variation, including impacts on voltage stability, ramping rates for other generators, reserve requirements, and potentially dumped output. The implementation used in this study accounts for three effects: increased reserve margin requirements with penetration, increased dumped output at higher penetration levels, and changes in the net load shape as a function of penetration.

As described in Kyle et al. (2007), the representation used here assumes that at low penetration levels the existing reserve margin requirements are sufficient to handle periods of low wind. As more wind generation is used, the wind farm must supply additional reserve margin, modeled primarily as an additional capital cost (assumed to be combustion turbines). The functional form of this requirement is given by:

$$AdditionalReserve = \sqrt{ReserveRequirement^2 + \sigma^2 W^2} - ReserveRequirement \quad (1)$$

where W is the amount of wind generation, ReserveRequirement is the normal amount of reserve required, and $\acute{o}^2$ is the variance in wind power output. As more wind power is used, the shape of the remaining load changes. This effect can be examined on a national basis by computing the residual load as a function of wind penetration using wind output on an hourly basis from a national wind dataset constructed by NREL[2]. For each hour of the year, the residual load is calculated as the national load curve minus wind generation. Wind penetration is measured as the fraction of the GCAM base load segment. As more wind is used the load curve shifts down.

US Residual Demand

As fn of gross wind generation (as fraction of GCAM baseload)

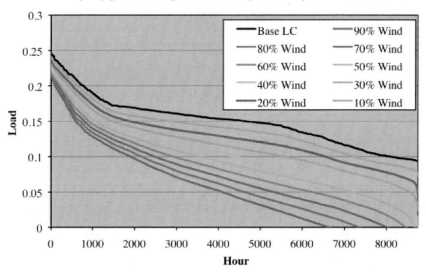

Figure 3. U.S. electricity demand after subtraction of wind generation.

From analysis of the residual load, several effects of increased wind penetration are apparent. First, the difference between the residual load and the original load curve occurs primarily in the base load segment as wind capacity increases (see *Figure* 3); this is demonstrated by the load curve somewhat uniformly shifting downwards with increasing wind penetration. This verifies that wind supplies primarily base load power. Second, at high wind penetration, wind output exceeds demand for some hours (particularly during nighttime). In Figure 3, this occurs for all cases above 50% wind, where the residual demand reaches zero for some hours of the year. Without storage or sufficient system flexibility, this power must be dumped. This excess output only occurs when wind is supplying greater than about 50% of base load generation. Third, as wind penetration increases, the reduction in base load demand is larger than the amount of wind generation by about 10%. The difference manifests primarily as an increase in demand in the sub-peak segment. The reason for this effect is that due to varying wind output at different hours and seasons, generation in the sub-peak segment must "fill in" to meet load. Therefore, this study represents this phenomenon as a shift in the load curve over time as more wind is used in the system.

The estimated dumped load (also known as curtailment) that occurs when wind generation exceeds demand is a lower limit to the amount of actual dumped load. Wind output would also need to be dumped if wind output were to exceed base load generation. This effect has been examined in the context of PV generation by Denholm and Margolis (2007a, b). Here, bounds are placed on this effect using the residual demand analysis shown above. Wind power might also be dumped due to transmission limitations. It might not be economic, for example, to build or upgrade transmission to handle relatively rare instances of maximum output from a remote wind farm. Dumped wind estimates from a variety of literature sources informed this exercise.

Wind Curtailment

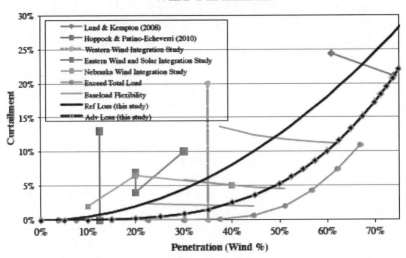

Figure 4. Wind curtailment: results from literature (including ranges from two studies) and present analysis and curtailment parameterization used in this study (black lines). Literature results shown are: Eastern Wind and Solar Integration Study (EnerNex Corporation 2010a), Nebraska Statewide Wind Integration Study (EnerNex Corporation 2010b), Western Wind and Solar Integration Study (GE Energy 2010), Hoppock, and Patino-Echeverri (2010), and Lund and Kempton (2008).

A graphical synthesis of the above analysis along with a summary of results from the literature is shown in Figure. Vertical lines indicate ranges found in a particular study (with the upper value generally representing an extreme value derived by assuming, for example, no new transmission construction).

Two loss curves are derived. Dumped power in the Reference Scenario is about 5% at 30% penetration and increases to 10% at 50% penetration. The Advanced Scenario assumes more advanced grid and load management, and has a lower rate of dumped wind, reaching 5% at 50% penetration.

1.3.2. Offshore Wind

Offshore wind power is structured similarly to onshore wind, with individual base load technologies implemented for each wind class. Offshore wind contains an additional level of detail, with technologies and associated resources for three ocean depth categories: shallow, transitional, and deep water. Wind resources are characterized by available area as a function of distance to grid. In the case of offshore wind, the distance provided is the distance to shore – the distance over which expensive (relative to land-based transmission) underwater cable must be constructed to provide offshore wind power to the onshore electric transmission grid.

The cost of offshore wind is a combination of exogenous capital costs, an exogenouslyestimated cost of building underwater transmission lines, and technology performance parameters. Offshore wind capital costs are the sum of turbine costs and the additional costs related to collecting and transmitting power to shore, adjusted for a transmission loss percentage assigned to each depth. Offshore transmission losses are assumed to range from 4% for shallow to 7% for deep water, based on the average distance to shore for each depth class. The performance parameters are generally the same as for onshore wind farms.

The wind speed input data used are from UNEP's Solar and Wind Energy Resource Assessment project, collected by NREL. The QuikSCAT satellite estimates annual wind speeds at a height of 10m by measuring ocean surface roughness. To estimate wind speed at 50m, power law exponents of 0.10 and 0.30 are used to adjust wind power and speed, respectively. Climatic conditions can cause the actual shear values to range from below 0.10 to above 0.15 for speed and 0.30 to 0.45 for power. These values also change seasonally.

NREL then averaged the data over the course of a five-year period (2000-2004) to obtain monthly averages of wind speed and power. Power is measured in W/m^2, given by:

$$Power = 0.5 \times \rho \times V_w^3 \qquad\qquad (2)$$

where P is the value of sea-level air density (1.225 kg/m^3), and V_w is wind speed. If there were fewer than four observations for a cell in a given month, the data are excluded. The satellite measurements may vary from those taken by an anemometer. Annual wind speed accuracy is estimated to be ± 1 m/s. This uncertainty can result in a change of up to two power classes and a greater total speed range of 5-10m/s. Instances of fewer observations and larger errors tend to occur in areas near coastlines and closer to shore. Note that the QuickSCAT satellite measures wind speed at two fixed times per day. A biased estimate could result if there is a strong diurnal wind pattern.

Three water depth classes are used. Shallow water depth is 0-30m; in this class, the turbine tower structure is similar to those of onshore wind, though more costly. Transitional water depth is 30-60m, where a more extensive tower is necessary. Deep water is greater than 60m, where floating turbine platforms must be used. Water depth data from NOAA (ETOPO1 dataset) are used for determining areas for the three depth classes.

In this exercise, areas within nine kilometers (five nautical miles) of shore are excluded. Also excluded are two-thirds of the area between 9-40 km from shore, and one-third of more distant areas. Depths exceeding 200m are also excluded from the data. A recent EEA study (Europe's onshore and offshore wind energy potential; EEA 2009) assumes even more stringent limits, excluding 96% of area between 0-10 km, 90% between 10-50 km, and 75% of the area beyond 50 km. The impact of more stringent exclusion values could be examined.

There are several fixed costs that comprise the total capital cost of offshore wind:

$$CapitalCost_{OffshoreWind} = Turbine + Collection\ \&\ Transmission + Foundation + BOS\ \&\ Fixed$$

(3)

The turbine cost is the cost of the turbine and tower structure, and is assumed to be 10% greater than land-based wind turbines. The collection cable cost assumes a wind farm turbine spacing of 630m and is based on the analysis of Green et al. (2007). Transmission cost is equal to the distance to shore multiplied by an assumed transmission cable cost, expressed in units of dollars/MW/km. Transmission to shore can us AC technology for distances of around 80-100km. For greater distances, DC transmission must be used. Transmission costs are based on the average of the three cases reported in Green et al. (2007). The foundation cost is the cost of the base of the wind turbine structure. It is assumed that the foundation cost is higher in deeper water because the platform must be taller and more structurally complex. Deep water turbines must utilize a floating platform technology. Foundation costs are based on a combination of estimates from NREL (Musial and Butterfield 2004) and project data reported by Greenacre et al. (2010). Note that the shallow water foundation costs reported in Greenacre et al. (2010) are much higher than those in Musial and Butterfield (2004). An intermediate value is used here. The remaining costs are balance-of-system (BOS) and fixed costs, which refer to costs other than the turbine, cabling, and transmission – such as labor, electricity, and operation permits.

While it is possible that off-shore wind may have different temporal characteristics than onshore wind, analysis of the supply-duration curve for off-shore wind generation on the U.S. east coast from Lund & Kempton (2008) finds a nearly identical behavior to the U.S. onshore data. Since a large portion of the U.S. wind resources is located off the east coast, we treat intermittency for offshore wind identically to onshore wind.

1.3.3. Concentrating Solar Thermal Power (CSP)

Concentrating Solar thermal Power (CSP) plants are implemented using the methodology described in Zhang et al. (2010). For simplicity, this study assumes that only the highest quality solar resources are used for CSP. The CSP resource is taken to be land with the best resources, taken from the resource categories used in Zhang et al. (2010). Irradiance and the number of nonoperational cloudy days were estimated for each sub-region.

CSP plants are assumed to be operated as hybrid systems with natural gas or biomass backup boilers, thereby providing dispatchable power. Three configurations of CSP plants are implemented – CSP without thermal storage supplying intermediate power, CSP plants with four hours of thermal storage supplying intermediate power, and CSP plants with ten hours of thermal storage supplying base load power. A comparison of CSP output by season with a California load curve, for example, found that CSP plants operating as intermediate segment generators are a very close match to the state's seasonal intermediate load demand.

All CSP plants are assumed to make use of the hybrid backup system to produce power on cloudy days when the solar field is not operational (excluding scheduled maintenance). CSP plants operating as intermediate load plants also must use backup power at some time periods as CSP penetration increases (Zhang et al. 2010). Because CSP capital costs are assumed to fall over time, and natural gas costs often increase in GCAM scenarios (particularly under a carbon policy), backup costs can become a significant portion of total CSP levelized costs. The explicit representation of this interaction results in a more accurate representation of the potential role of CSP. CSP plants serving intermediate load also are subject to increasing dumped load as penetration increases, although for CSP plants with thermal storage this is small unless CSP is supplying a large fraction of intermediate power.

1.3.4. Photovoltaics

Photovoltaic (PV) technologies are implemented in two forms: as central PV generation plants and as distributed rooftop PV technologies. Rooftop PV is implemented in the building electricity distribution sector, where it competes with grid electricity to supply building electric loads using the supply curves for each sub-region from Denholm and Margolis (2008). Rooftop PV is more expensive than central PV installations, due to economies of scale and lower installation costs for central systems. Rooftop systems are assumed to compete against delivered electricity, which has a higher price as compared to the average generation cost seen by a central PV plant.

Central PV plants are implemented assuming an average irradiance value for each sub-region assuming systems tilted at latitude. The electric generation module includes representations of central PV with and without a dedicated electricity storage system.

A comparison of PV output with the California load curve by season, taken as typical of sunny regions, as well as analysis of the results from Denholm and Margolis (2007a), finds that while the largest portion of PV contributes to intermediate load, a portion of PV output offsets sub-peak and peak segments. Therefore, this study assumes that PV without storage contributes 25% of its power to the sub-peak load segment. PV with storage also contributes to peak and base load generation.

Intermittency is a substantial issue for PV power. This treatment of intermittency for PV draws from the analysis of Denholm and Margolis (2007a, b), who examine the impact of PV penetration in the Texas region. They find that dumped power becomes significant at penetration levels as low as 10% for PV systems without storage. PV systems augmented by electricity storage (or flexible generation response) can supply a greater fraction of load without dumped power. Therefore, a dumped power parameterization based on the Denholm and Margolis analysis is used for central PV systems. A backup requirement is also implemented such that, at high PV penetration levels, reliable power is provided during cloudy days.

1.3.5. Geothermal

Geothermal technologies are assumed to be base load technologies with no intermittency. Three geothermal technologies are implemented î hydrothermal, near-field hydrothermal engineered geothermal, and deep engineered geothermal (EGS). Geothermal technologies are implemented in the GCAM as described in Hannam et al. (2009), using updated resource and cost estimates from Augustine et al. (2010).

1.3.6. Solar Hot Water

Residential solar hot water is implemented as described in Smith et al. (2010), drawing on the methodology of Christensen and Barker (1998). Because solar hot water heaters can have natural gas or electric back-up, they can essentially be treated as high-efficiency natural gas and electric hot water technologies in residential buildings. As such, they are assigned an efficiency and non-energy service cost, representing levelized capital and O&M costs of the systems. The "efficiency" used by the GCAM is technically an I-O (input-output) coefficient that represents the amount of energy input needed for a unit of service output (Smith et al. 2010). Solar hot water heaters' I-O coefficients are given by:

$$C_{I-O} = \left(\frac{E_{SHW}^{aux}}{E_{conv}} \right) \div \varepsilon_{conv}$$

(4)

where the I-O coefficient (CI-O) is equal to the ratio of the auxiliary energy consumed by a solar hot water system to the energy consumed by a conventional hot water system (Econv), divided by the system efficiency of a conventional water heater (Econv). These variables are specific to each sub-region, incorporating incident solar energy data from NASA, temperature differences between water mains and hot water tanks, and residential hot water draw (Smith et al. 2010).

1.3.7. Renewable Capital Costs

The capital cost assumptions for renewable technologies are shown in the table below. Capital costs for most electric generation technologies have increased in recent years. The renewable costs below are based on literature reviews and recent assessments. Because the core GCAM model capital costs for non-renewable technologies are baselined to earlier, and lower, cost data the cost figures below were reduced by 20 – 25% in order to produce a level competition within the model. In general, the recent increase in power plant costs overall leads to substantial uncertainty in future costs. It is not clear, for example, how much of these recent cost increases might be temporary, or how much these reflect fundamental shifts in specific technologies or global cost structures.

Table 1. Capital Cost Assumptions

		Reference Tech		Advanced Tech		
Technology	2005	2050	2095	2050	2095	Units
Onshore Wind (Class 7)	1,278	1,099	1,045	720	657	2005$/kW
Offshore Wind (Shallow Average - Class 7)	2,078	1,788	1,699	1,083	964	2005$/kW
CSP - Intermedite with Th Storage	4,578	3,476	2,621	2,437	1,587	2005$/kW
CSP - Baseload (with Th Storage)	5,410	4,092	3,070	2,813	1,800	2005$/kW
Central PV	6,905	2,479	1,766	1,387	951	2005$/kWp dc
Geothermal (Ca - hydrothermal)	2,556	2,063	1,836	2,063	1,836	2005$/kW
Geothermal (WECC North EGS)	-	-	-	4,076	4,076	2005$/kW

2. RESULTS

2.1. Scenario Design

The goal of this study was to examine the role of U.S. renewable generation in the GCAMRE model. Three dimensions were examined: technology cost and availability, availability of long-distance transmission, and stringency of a climate policy target. The scenarios considered are shown in the table below.

Table 2. Roles of Renewable Generation in the GCAM-RE Model

Renewable Technology	Other Electric	Transmission	Policy Target (W/m^2)
Reference Advanced	Reference No CCS/No New Nuclear Sites	Regional Intermittency Mitigation National Renew Grid	None 5.8 4.5 3.7 3.0

In the advanced technology scenarios, costs for renewable technologies were assumed to fall faster in the future as compared to the reference scenario. Wind dumped load was also assumed to occur at lower levels (*Figure 4*), and engineered geothermal generation is assumed to be available.

Two scenarios for other electric generation technologies are examined. In the first, there are no exogenous limits on any electric generation technology. A full suite of technologies compete, as appropriate, to supply electricity in the four generation segments.

In the second scenario, Nuclear generation is limited to current levels and carbon-dioxide capture with geologic storage (CCS) is assumed to be unavailable. These can be considered bounding scenarios, with the first scenario perhaps overly optimistic, since not every technology will be suitable, or acceptable, in every region, and the second overly pessimistic.

The range of climate policy targets ranges from no climate policy, which is the reference case, to a very aggressive policy that meets a radiative forcing target of 3.0 W/m2 by the end of the century. The policy cases are implemented in two stages. First, the GCAM-RE model is run with a global carbon tax which results in the end of century radiative forcing as indicated in the scenario name.

Carbon dioxide emissions in the United States were then constrained to follow identical pathways for the alternative renewable technology and transmission cases. The global carbon price path associated with each scenario is applied to the rest of the world. This process was repeated for the set of scenarios without nuclear or CCS. These scenarios result in higher global and U.S. carbon prices. Carbon dioxide concentrations and U.S. emissions are shown in *Figure 5*.

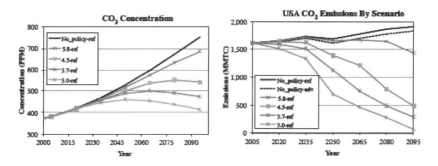

Figure 5. Carbon dioxide concentrations and carbon dioxide emissions pathways for the U.S.

Three transmission cases are considered. In the regional transmission case all wind, solar, and geothermal generation must be consumed within each U.S. sub-region. The national grid scenario adds an additional option whereby renewable generation can be consumed in any U.S. sub-region, although with an additional capital charge and transmission loss. The intermittency mitigation scenario examines the impact of changing the modeled constraints due to intermittency. The alternative transmission scenarios will be further described in the results section below.

2.2. Renewable Generation

A common finding in IA models, including GCAM, is that the implementation of a comprehensive carbon policy spurs an increase in electricity demand. As electric generation decarbonizes, this spurs a shift in end-use sectors away from direct use of fossil fuels, which are now subject to a carbon price, toward electricity. This is shown in Figure 6, where electricity generation increases to slightly over 35 EJ by the 2095.

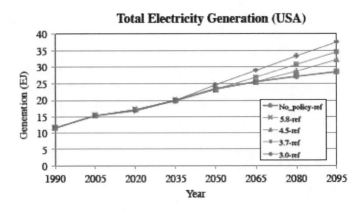

Figure 6. Total U.S. electricity generation for the reference case scenarios.

The cost of electricity generation increases over time, relative to the reference case due to a carbon policy. Costs for the reference case and the 4.5 W/m2 scenario are shown in *Figure 7*. While under the reference case generation costs are fairly stable, costs are 20-40% higher under a carbon policy, relative to the reference case. Peak electricity costs increase the most, while costs for base load and intermediate generation show the smallest increase. This is because there are many more low-carbon options for base load and intermediate generation, while there are relatively limited options for the more flexible generation needed to provide peak and sub-peak power.

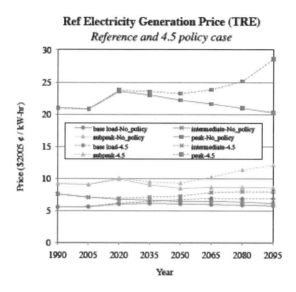

Figure 7. Electricity generation costs in the TRE (Texas) region under the reference and 4.5 W/m2 climate policy scenarios.

Note that the cost to the consumer will be a weighted average of the costs for each segment. Changes in electric demand that might reduce the need for peak and sub-peak power was not considered in this study.

Total renewable generation is shown in *Figure 8*. Renewable generation increases over time in all scenarios due to both increased electric demand in general and an assumed decrease in renewable technology costs. Renewable generation is higher if the renewable costs are assumed to be smaller, which is the case in the advanced scenario. Renewable generation also increases as the radiative forcing target is lowered. The fraction of U.S. electricity provided by renewables increases over time. The flattening of renewable percentage in some cases is due to model dynamics, particularly competition from other electric generation sources, but also the extent to which the generation profile of renewables can contribute to different generation segments. There is no set limit to renewable penetration in these scenarios. Generation fractions from the no new nuclear or CCS scenarios have generation fractions that are much higher.

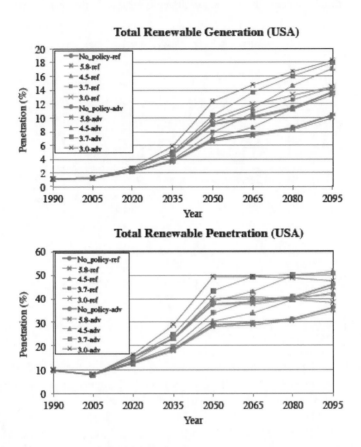

Figure 8. Total U.S. renewable generation, including biomass and hydro power, generation. Absolute value (top) and percentage of total generation (bottom) are shown. Results are shown for reference (thick lines) and policy scenarios, as well as for reference technology (red) and advanced technology (blue) assumptions.

U.S. electricity production by technology is shown in *Figure 9* for the 4.5 W/m2 policy scenario with advanced renewable assumptions. Biomass, CSP, on and off-shore wind, and PV make large contribution to U.S. electric generation under this scenario. Biomass generation is particularly important, as biomass coupled with CCS, can provide net negative emissions. Wind, solar, and geothermal supply 33% of U.S. electric generation in 2095 for this scenario.

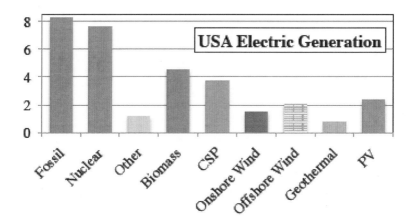

Figure 9. U.S. generation by type in 2095 under the 4.5 W/m2 policy case, with advanced assumptions for renewable technologies characteristics.

Figure 10 shows generation by sub-region for the same scenario. The regional generation mix varies considerably. Note that no regional differences in technology preferences or mandates were assumed. The differences shown here are solely due to the different assumptions for regional costs and resources of solar, wind, and geothermal technologies. The relative role of different renewable technologies reflects the regional distribution of renewable resources.

The renewable percentage is lowest in the southeast, RFC_West, and RFC_East regions (see Figure 10). The renewable fraction in the mid-west is around 40%, while on the west and southwest, the renewable fraction is around 50%.

CSP solar technologies play the largest role in the west and south west. Offshore wind is largest along the east coast, and on-shore wind plays a larger role in the mid-west.

Under a carbon dioxide emissions constraint, regions with abundant renewable resources preferentially use those resources. Those regions without good quality renewable resources rely more heavily on fossil, nuclear, or biomass. The ability, in principle at least, to build fossil, nuclear, or biomass plants in any region is a key flexibility.

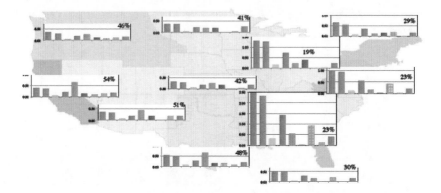

Figure 10. U.S. generation by type and sub-region in 2095 under the 4.5 W/m2 policy case, with advanced assumptions for renewable technologies characteristics. For legend see Figure 9.The percentage of wind, solar, and geothermal is shown for each sub-region.

2.3. Impact of National Transmission

In the national grid scenarios, any sub-region can, in addition to regional generation, also consume renewable generation from elsewhere in the United States through a national grid, paying an additional ~ 1 cents/kWhr along with an additional 5% transmission loss. Figure 11 shows the impact of a national grid, for the 4.5 W/m2 scenario assuming advanced renewable technologies. There is an increase in the consumption of CSP and on-shore wind, as these are two of the most competitive technologies, and a decrease in fossil and nuclear generation. Note, however, that these shifts are relatively small compared to total U.S. electricity generation of ~ 32 EJ in 2095 in this scenario.

If no new nuclear or CCS is assumed, then the shifts are twice as large for on-shore wind and 3 times as large for CSP.

The intermittency mitigation scenario had very little impact on the results. In this scenario intermittency effects, such as wind and PC curtailment and reserve margin requirements, where evaluated against national renewable penetration levels instead of regional levels. The low impact of this change in assumption may mean that a modeling simplification may be reasonable whereby only national markets instead of multiple regional markets are tracked.

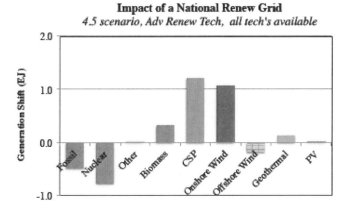

Figure 11. Difference in generation between the regional transmission scenario and the national grid scenario. A positive value means that more generation of that type is used in the national grid scenario.

2.4. Renewable Generation and Climate Policy Costs

The carbon price for several carbon policy cases is shown in *Figure 12*. In the original scenario set-up global carbon prices increases at a constant rate over time. When the U.S. carbon constraints are applied using the detailed U.S. regional model, the same basic character is seen. Under the weakest carbon constraint, the reference case using the detailed renewable model is below the carbon constraint until 2065 due to a larger penetration of renewables as compared to the core GCAM model. Renewable generation in this scenario is sufficient to keep reference case emissions below the specified carbon constraint path. The 4.5 W/m^2 case meets the constraint in 2020, with the carbon price increasing to 500 $/tC in 2095. The two most stringent scenarios (3.0 W/m^2 and 3.0 W/m^2) reach carbon prices of 1,000 and 2,000 $/tC by 2095.

More important than the absolute values are the changes in carbon prices under a change in assumptions. The table below (rounded to 5%) shows how the carbon prices change, relative to the reference technology scenario, for several alternatives: advanced renewable technologies, no CCS or New Nuclear, and the national grid scenario. An important general tend is that renewable assumptions only impact the results when the carbon prices are under about $100/tC. Above this level the electricity generation is largely decarbonized and carbon prices depend on the ability of end-use sectors to reduce carbon emissions.

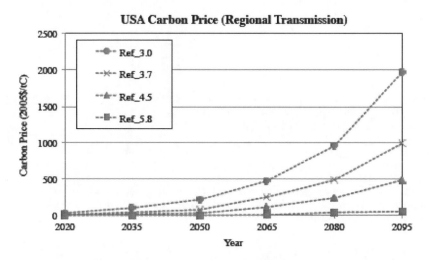

Figure 12. USA carbon price for several climate policy targets under reference technology and regional transmission assumptions.

In the early stages of a carbon policy, renewable generation assumptions have a significant impact on carbon prices. More favorable assumptions for renewable technologies across the board, as in the "advanced" renewable technology scenario, lower carbon prices significantly in early years. Note that the relative change in the very first year of a policy can be large due to a low carbon price in that first year. The carbon price in the second policy period is 15-40% lower under the more favorable "advanced" case assumptions for renewable generation.

Removing the option of building new nuclear plants and the use of carbon-dioxide capture and geologic storage (CCS) increases costs at all periods.

Cost increases are particularly large in later years as CCS is a critical technology as it allows emissions reductions in sectors such as refining, and also allows net negative emissions when coupled with combustion of biomass. In early years, the impact of the noNuc-CCS assumption has a somewhat larger relative influence, although of opposite sign, than the advanced renewable assumptions.

Table 3. Carbon Price Change From Reference Technology with Regional Transmission Scenario

	2020	2035	2050	2065	2080	2095
Adv 5.8	-	-	-	-100%	-25%	-5%
Adv 4.5	-	-70%	-40%	-5%	0%	0%
Adv 3.7	-60%	-15%	-10%	0%	0%	0%
Adv 3.0	-20%	-5%	0%	0%	0%	0%
Ref 5.8 No CCS or NewNuc	-	-	-	440%	90%	100%
Ref 4.5 No CCS or NewNuc	-	55%	80%	80%	155%	210%
Ref 3.7 No CCS or NewNuc	-10%	35%	85%	160%	215%	155%
Ref 3.0 No CCS or NewNuc	0%	40%	175%	225%	210%	*
AdvRenew-NoNucCCS 5.8	-	-	-	45%	45%	70%
AdvRenew-NoNucCCS 4.5	-	-40%	5%	55%	145%	210%
AdvRenew-NoNucCCS 3.7	-85%	10%	50%	150%	215%	160%
AdvRenew-NoNucCCS 3.0	-20%	25%	165%	225%	215%	*
Natl Grid 5.8	-	-	-	-55%	-5%	0%
Natl Grid 4.5	-	-30%	-15%	0%	0%	0%
Natl Grid 3.7	-35%	-5%	-5%	0%	0%	0%
Natl Grid 3.0	-10%	0%	0%	0%	0%	0%

* Note that the model did not solve the 3.0 case without CCS in 2095 because it was not possible to produce the net negative emissions required in this scenario.

Also shown in the table is the combination of advanced renewable assumptions with the assumption of no-new nuclear and no CCS. In the early stages of a climate policy, more favorable renewable assumptions can offset less favorable assumptions for nuclear and CCS. As policy emission targets get lower, however, the lack of CCS options increases costs considerably. This is due to the importance of CCS in lowering emissions in sectors such as refining and cement (Luckow et al. 2010), where renewable options were not assumed to be available in these scenarios.

Finally, the presence of a national renewable grid also reduces costs, although by 5-15% in the second policy period. Allowing renewable energy production to be used in any sub-region, even with a cost penalty, allows sub-regions with limited renewable resources to use, for example, wind from the mid-west or solar from the southwest. Note that such a national grid might have additional benefits in terms of integrating renewables, such as allowing better matching between load and demand on a continental scale, but these benefits could not be evaluated in this framework.

CONCLUSION

This project has demonstrated the ability of the GCAM integrated assessment model to simulate renewable electricity generation at a sub-regional level in the United States. Regional renewable resources differ greatly across the United States and adding the ability to consider 12 U.S. sub-regions allows analysis of the impact of these differences. No overall constraints on renewable generation were applied, with the impact of intermittency addressed for each technology.

Renewable generation, including hydro and biomass, comprise a large portion of U.S. electricity supply in these scenarios, supplying 35-50% of electric generation by the end of the century, even in scenarios allowing competition with a full suite of electric generation technologies. Renewable penetration was highest in the west/south west, and lowest in the east/south-east. Use of renewable generation increases under an economy-wide carbon constraint, and also if future costs of renewable technologies are assumed to be lower and a more flexible electricity system is assumed.

Wind and CSP thermal power are the largest renewable sources. Offshore wind makes a substantial contribution, particularly along the east coast, in the advanced scenario where offshore wind farm transmission and construction costs are assumed to fall more rapidly.

The availability of a national transmission grid can further reduce costs, and changes the generation mix. The presence of a national transmission grid allows greater use of onshore wind and CSP in sub-regions where renewable resources are more limited.

A general finding of integrated energy models is that the electric generation system largely de-carbonizes in the early stages of a comprehensive carbon pricing policy (Clarke et al. 2007). Consistent with this, we find that renewable assumptions have the most impact in the early years of a climate policy, which is where most of the transformation of the electricity system occurs.

We demonstrate that renewable electricity generation can provide important flexibility in the lowering costs in the early stages of a climate policy.

Climate policy costs in latter years, when emissions targets are lower, did not depend on renewable assumptions, but instead on assumptions for the remainder of the energy system, such as low-carbon options in the refining sector and the potential for electric vehicles in the transportation system.

ACKNOWLEDGMENTS

Primary funding for this project was from the U.S. Department of Energy's Office of Energy Efficiency and Renewable Energy, with additional support from the Global Technology Strategy Project and the California Energy Commission. Some methodological descriptions in this report are shared with other reports that use the same model. The authors would like to thank Patrick Luckow for helpful comments on the draft report.

REFERENCES

Augustine, C., K.R. Young, and A. Anderson (2010). Updated U.S. Geothermal Supply Curve. Golden, CO: National Renewable Energy Laboratory, NREL/CP-6A247458.

Brenkert, A., Smith, S., Kim, S., Pitcher, H., 2003. Model Documentation for the MiniCAM. PNNL-14337, Pacific Northwest National Laboratory, Richland WA.http://www.pnl.gov/main/publications/external/ technical_reports/PNNL-14337.pdf.

Christensen, C. and G. Barker (1998). "Annual System Efficiencies for Solar Water Heating." Proc. ASES, ASES, Boulder, CO.

Clarke, J,F and Edmonds, J.A., 1993. Modeling energy technologies in a competitive market. *Energy Economics* 15, 123-129.

Clarke, L., J. Edmonds, H. Jacoby, H. Pitcher, J. Reilly, and R. Richels (2007). Scenarios of Greenhouse Gas Emissions and Atmospheric Concentrations. Sub-report 2.1A of Synthesis and Assessment Product 2.1 by the U.S. Climate Change Science Program and the Subcommittee on Global Change Research. Washington, D.C.: U.S. Department of Energy, Office of Biological & Environmental Research.

Clarke, L., Weyant, J., Edmonds, J., 2008b. On sources of technological change: what do the models assume? Energy Economics 30, 409-424.

Clarke, L., Wise, M.A., Edmonds, J., Placet, M., Kyle, P., Calvin, K., Kim, S., Smith, S., 2008b. CO2 Emissions Mitigation and Technological Advance: an Updated Analysis of Advanced Technology Scenarios (Scenarios Updated January 2009). PNNL-18075, Pacific Northwest National Laboratory, Richland, WA. http://www.pnl.gov/science/pdf/ PNNL18075.pdf.

Denholm, P. and R. Margolis (2007a). "Evaluating the limits of solar photovoltaics (PV) in traditional electric power systems." *Energy Policy* 35: 2852–2861.

Denholm, P. and R. Margolis (2007b). "Evaluating the limits of solar photovoltaics (PV) in electric power systems utilizing energy storage and other enabling technologies." *Energy Policy* 35: 4424–4433.

Denholm, P., and R. Margolis (2008). Supply curves for rooftop PV-generated electricity for the United States. Golden, CO: National Renewable Energy Laboratory, NREL/TP-6AO44073.

EnerNex Corporation. 2010a. Eastern Wind and Solar Integration Study. NREL/SR-550- 47078. Golden, CO: National Renewable Energy Laboratory.

EnerNex Corporation. 2010b. Nebraska Statewide Wind Integration Study. NREL/SR-550- 47519. Golden, CO: National Renewable Energy Laboratory.

GE Energy. 2010. Western Wind and Solar Integration Study. NREL/SR-550-47434. Golden, CO: National Renewable Energy Laboratory.

Green, J., A. Bowen, L.J. Fingersh, and Y. Wan (2007). Electrical Collection and Transmission Systems for Offshore Wind Power. Golden, CO: National Renewable Energy Laboratory, NREL/CP-500-41135.

Greenacre, P., R. Gross, and P. Heptonstall (2010). Great Expectations: The cost of offshore.

wind in UK waters – Understanding the past and projecting the future. London: UK Energy Research Centre, ISBN 1903144094.

Hannam, P., P. Kyle, and S.J. Smith (2009). Global Deployment of Geothermal Energy Using a New Characterization in GCAM 1.0. College Park, MD: Battelle Memorial Institute, PNNL-19231.

Hoogwijk, M., de Vries, B., Turkenburg, W., 2004. Assessment of the global and regional geographical, technical and economic potential of onshore wind energy. Energy Economics 26, 889-919.

Hoppock, D.C. and D. Patino-Echeverri, 2010. Cost of Wind Energy: Comparing Distant Wind Resources to Local Resources in the Midwestern United States. Environmental Science & Technology 40, 8758-8765.

Kim, S., Edmonds, J., Lurz, J., Smith, S., Wise, M., 2006. The Object-oriented Energy Climate Technology Systems (ObjECTS) framework and hybrid modeling of transportation in the MiniCAM long-term, global integrated assessment model. The Energy Journal, Special Issue: Hybrid Modeling of Energy-Environment Policies: Reconciling Bottom-up and Top-down, 63-91.

Kyle, P., S.J. Smith, M.A. Wise, J.P. Lurz, and D. Barrie (2007). Long-Term Modeling of Wind Energy in the United States. College Park, MD: Battelle Memorial Institute, PNNL16316.

Luckow P, MA Wise, JJ Dooley, and SH Kim. 2010. "Large-Scale Utilization of Biomass Energy and Carbon Dioxide Capture and Storage in the Transport and Electricity Sectors under Stringent CO2 Concentration Limit Scenarios." International Journal of Greenhouse Gas Control 4(5):865-877. doi:10.1016/j.ijggc.2010.06.002.

Lund, H. and W. Kempton, 2008. Integration of renewable energy into the transport and electricity sectors through V2G. Energy Policy, doi:10.1016/j.enpol.2008.06.007.

Musial, W. and S. Butterfield (2004). Future for Offshore Wind in the United States. Golden, CO: National Renewable Energy Laboratory, NREL/CP-500-36313.

U.S. DOE, 2008. 20 percent Wind Energy by 2020, Increasing Wind Energy's Contribution to U.S. Electricity Supply. DOE/GO-102008-2567. Office of Energy Efficiency and Renewable Energy, U.S. Department of Energy, Washington, D.C. http://www1.eere.energy.gov/windandhydro/pdfs/41869.pdf.

Wise, M.A., and S.J. Smith, 2007. Integrating Renewable Electricity, Electricity Demand, and Electricity Storage: A New Approach for Modeling the Electricity Sector in ObjECTS PNNL-16500.

Zhang, Y., and S.J. Smith (2007). *Long-Term Modeling of Solar Energy: Analysis of PV and CSP Technologies*. PNNL-16727.

Zhang, Y., S.J. Smith, G.P. Kyle, and P.W. Stackhouse Jr. (2010). *"Modeling the Potential for Thermal Concentrating Solar Power Technologies"*. Energy Policy 38: 7884–7897.

INDEX

A

access, vii, 2, 5, 6, 9, 12, 13, 18, 19, 20, 23, 27, 30, 31, 32, 33, 39, 53, 55, 57, 65, 104, 105, 116, 118, 122, 124, 129, 137, 138, 139, 141, 142, 150, 152
accounting, 20
actual output, 86
adaptation, 102
advancements, 38
advertisements, 122
aesthetic(s), 17, 78, 122
agencies, 9, 14, 94, 154
agriculture, 164
architect, 58
articulation, 134
assessment, 10, 34, 35, 36, 53, 59, 64, 84, 118, 149, 155, 164, 186, 188
assets, 56, 57, 96, 99, 105
atmosphere, 38
Australia, vii, 1, 2, 3, 4, 5, 6, 7, 8, 18, 23, 24, 26, 27, 37, 38, 49, 50, 53, 54, 55, 58, 59, 61, 62, 63, 64, 65, 66, 68, 69, 70, 71, 150, 156, 157, 160
Austria, 107
authority(s), 10, 16, 18, 119, 143, 147, 157
automate, 27
autonomous communities, 130
awareness, 124

B

barriers, 9, 30, 57, 92, 128, 159
base, vii, 2, 7, 8, 21, 44, 93, 103, 136, 165, 166, 168, 169, 170, 171, 172, 173, 174, 175, 178
batteries, 28
Beijing, 114
Belgium, 110
benefits, 16, 26, 29, 40, 41, 43, 53, 55, 57, 58, 59, 75, 97, 102, 105, 112, 118, 119, 122, 124, 132, 134, 147, 148, 150, 152, 162, 185
bias, 68
Bilateral, 134
biomass, 49, 50, 74, 93, 97, 103, 104, 117, 173, 180, 181, 184, 186
boilers, 28, 173
bonuses, 39
BOS, 173
bounds, 170
Brazil, 156, 158
Britain, 122
businesses, 160

C

cables, 74, 75, 76
campaigns, 59
capacity building, 129, 139, 156

D

E

F